Thirst for Independence

The San Diego Water Story

Dan Walker

Sunbelt Publications
San Diego, California

Thirst for Independence
Sunbelt Publications, Inc.
Copyright © 2004 by the author
All rights reserved. First edition 2004

Edited by Lowell Lindsay
Copy editing by Laurie Gibson
Book design by W. G. Hample & Associates
Cover design by Phil Farquharson
Project management by Jennifer Redmond
Printed in the United States of America

Sunbelt Publications, Inc.
P.O. Box 191126
San Diego, CA 92159-1126
(619) 258-4911, fax: (619) 258-4916
www.sunbeltbooks.com
mail@sunbeltpub.com

08 07 06 05 04   5 4 3 2 1

Library of Congress Cataloging-in-Publication Data

Walker, Daniel, 1922-
        Thirst for independence : the San Diego water story / by Dan
        Walker.— 1st ed.
                p. cm. — (Adventures in cultural and natural history)
        ISBN 0-932653-62-6
        1. Water-supply—California—San Diego County—History. I. Title.
        II. Series.

TD224.C3W348 2004
363.6'1'0979498—dc22

                                                    2004014233

Cover photos, by permission:
Front cover — Bill White, Old Mission Dam, Mission Trails Regional Park
Front inset — Kim Thorner, Olivenhain Municipal Water District
Back cover — Phil Farquharson, personal collection

Dedication —

To Lily,

who made it all possible.

FOREWORD

"Whiskey is for drinking; water is for fighting over." These words, famously attributed to Mark Twain, describe many a water development scenario in the arid southwest United States. The San Diego region is no exception. With a sparse average annual rainfall that approaches true desert conditions and a burgeoning population that would require many times that amount for self-sufficiency, the region has long relied upon imported water. Local ground and surface water supplies maintained San Diego's water independence until the arrival of Colorado River water in 1946 and northern California water in 1978. For the last sixty years, the San Diego County Water Authority (SDCWA), one of over two dozen agencies under the umbrella of the Los Angeles-based Metropolitan Water District (MWD), has managed water importation into the county.

On July 23, 2004 in Denver, Secretary of the Interior Gale Norton described our current drought as "possibly the worst in 500 years," based on U.S. Geological Survey data and analyses. This is with particular reference to the parched Colorado River basin upon which San Diego is heavily dependent, far more so than the rest of southern California (which obtains a much larger percentage of its water from the Owens Valley and northern California). This condition lends ever-increased urgency "to improve the region's water reliability by diversifying our future water supplies and reducing our overdependence on MWD" per a SDCWA news release on July 14, 2004. SDCWA continues to slake San Diego's thirst through innovative water transfer agreements, enhancement of recycling and conservation measures, and desalination research and development. After all, as former county hydrologist John Peterson notes, "This isn't Arizona. We have a very big pond of water to the west."

Sunbelt's mission is to produce publications about the natural history and cultural heritage of the Californias. Our books address not only nature, but man-made effects upon our natural environment — city planning (*San Diego: An Introduction to the Region*), land use and environmental legislation (*California Desert Miracle*), learning about the land (*The Rise and Fall of San Diego*), and now resource management with *Thirst for Independence: The San Diego Water Story*. Culture, they say, is everything around us that was not created by nature, so this is another story about the land and its people — people who needed water, those who brought water, some who stole it, and a few who even claim to own it.

This work, by former Illinois governor Dan Walker, offers historical background, geopolitical overview, and engineering perspective to the water issues facing San Diego today and tomorrow. We hope this timely book will enable readers to understand and interpret today's controversial water headlines, as well as the politics and bureaucracy that decide who in the Golden State gets how much — and from where — of this precious natural resource.

Lowell Lindsay, Publisher

CONTENTS

Foreword   iv

List of Maps, Charts, and Tables   vi

Preface and Acknowledgments   vii

Introduction   ix

Chronology of Major Events — San Diego Water Story   x

Chapter 1   The Essential Background   1

Chapter 2   Dam Builders and Water Hustlers   11

Chapter 3   The Rape of Owens Valley   27

Chapter 4   Moving the Colorado River West   31

Chapter 5   Fateful Decisions for San Diego   49

Chapter 6   A Constantly Troubled Marriage   61

Chapter 7   Water Distribution in San Diego County   75

Chapter 8   Preparing for Emergencies and Disasters   93

Chapter 9   Always Seeking New Water   99

Chapter 10 The Ambitious Ag-Urban Water Transfers   115

Chapter 11 Journey to an Epochal Agreement   129

Conclusion   153

Selected References   157

Index   159

Recommended Reading   164

MAPS, CHARTS, AND TABLES

San Diego Aqueduct 54

California aqueducts 57

California Aqueduct East Branch 58

MWD member agencies 62

MWD voting rights 64

SDCWA member agencies 76

SDCWA member agency voting rights 77

SDCWA pipelines in San Diego County 80

San Diego County usage of local versus imported water, 1997-2001 83

Delivery of water in City of San Diego 85

San Diego City water reservoirs 87

Remaining SDCWA water districts 92

Priorities to Colorado River water 100

Reservoirs in San Diego County 105

Projected water conservation savings (SDCWA) 107

Desalination schematic 111

Estimated water demand (SDCWA) 113

Projected local water supplies (SDCWA) 113

Preferential MWD water rights 114

SDCWA water supply-demand projection 126

SDCWA water supply-demand projection 143

SDCWA water supplies, 2003-2020 150

PREFACE AND ACKNOWLEDGMENTS

This account of San Diego County's past, present, and future water problems is written more for those who live and work in the county than for historians or scholars. I have tried to include enough history to understand the present troublesome situation, enough engineering to describe San Diego's dependence on far-away rivers, and enough explanation to portray how our precious water is received, stored, distributed, and protected. And finally, enough analysis to highlight the current, serious water supply issues that confront San Diego; issues that have too often, regrettably, escaped the attention of most people in the county.

I am indebted to the officials and staff of San Diego's water agencies, particularly the San Diego County Water Authority (SDCWA), who have been so generous with their time and assistance. All who helped are too numerous to name, but some in addition to those quoted in the text deserve special mention. At SDCWA, Ivan Golakoff, who was supervising his own department's history of the agency, still took considerable time to help. Chief Engineer John Economides supplied expert professional aid and his assistant, Principal Engineer Ken Steele, helped on numerous occasions.

James Taylor, SDCWA Assistant General Counsel, was invaluable. Without him, I never could have understood all the legal and practical ramifications of the critical Imperial Valley water transfer arrangement and he was always willing to take time to answer tough questions. Jack Papp, Public Affairs Director, promptly and uncomplainingly responded to many requests for information and assistance.

Katherine Auld Breese of the Helix Water District provided critical historical insight and also graciously made available Harry Griffen's unpublished 1200-page *Harry Griffen Manuscript,* on file at the Helix Water District archives, that describes his many years of involvement as a "Mr. Water" for San Diego County, a source for invaluable historical detail, particularly on the relationship between SDCWA and Metropolitan Water District (MWD).

Extremely helpful was Linden Burzell's unpublished *Memoirs* based on his years of service on the SDCWA board of directors. At the City of San Diego Water Department, Director Larry Gardner patiently answered numerous questions and made his staff constantly available. And at MWD, Rob Hallwachs of the public affairs staff was always cheerful and ready to respond to my requests.

As independent commentators and distinguished professors, both Philip Pryde of San Diego State University and Steve Erie of the University of California at San Diego gave generously of their time and thoughts. Distinguished water attorney Paul Engstrand made many insightful comments on drafts of the manuscript and also helped by lending his copy of the recorded oral history of water attorney William Jennings (original manuscript lodged at University of California at Los Angeles library) which supplied much critical information about the past relations of SDCWA and MWD.

Colonel Ed Fletcher's unpublished *Memoirs* provided many helpful historical insights and stories. I obtained his memoirs and much other useful information from the San Diego Historical Society archives (a treasure trove for those searching into San Diego's past). And, the book would never have made it into print without the delightfully talented Jennifer Redmond at Sunbelt Publications, my editor *non pariel* and always-smiling *obertaskmeister*.

Any errors made in digesting and writing from the material and information provided by these and many other sources are mine alone.

Dan Walker

INTRODUCTION

Drinking water (only 1% of the world's water) has always been critical in America's West but people there, like people everywhere, tend to forget that it is the stuff of life and should be counted among nature's most precious gifts. Marc Reisner's groundbreaking book *Cadillac Desert* described dramatically how Americans have used, abused, protected, controlled, fought over, and died for water in the western states of the nation.

This book, focusing on just one part of the West, is a story of nature quiet and nature on a rampage, of successes and failures, of water used for good and wasted for bad. Here are tales of human avarice, of some politicians providing leadership while some were degraded by bribery and corruption, of newspapers often leading the community but sometimes seeming to just help the rich get richer, of men and women of all ilk, some of whom supplied leadership, some of whom were content to just stuff their pockets, many who pitched in with help when needed, and too many who apathetically watched the water crises come and go with seeming indifference. Finally, this book tells of governmental agencies, officials, and employees who strive to do their best while coping with difficult problems.

For many decades, San Diego was a sleepy little town tucked away in the southwestern corner of the nation, known for its beautiful, landlocked harbor, admired for its world-class zoo, and rated as a good liberty town for visiting sailors. Not until World War II did San Diego commence its real growth, becoming a center for military production and a premier naval port. San Diego finally achieved its saltwater destiny as ships loaded with men and women, airplanes, and equipment sailed across the oceans to combat destinations on the far side of the world.

This book tells another water story. From hand-carried water buckets to the largest dam-building spree in the world, to huge reservoirs, aqueducts and canals, San Diego has moved aggressively to build a modern freshwater storage, transfer, and distribution system serving millions of people, always searching for a secure long-term water supply.

In the beginning, San Diego's thirst was for freedom from the constraints imposed by nature, a search welcomed by all who lived and worked in this arid land. In the end, another dimension has been added. San Diego has become involved in a quest for local governmental independence to find ways to ensure an adequate, reliable water supply for the future.

CHRONOLOGY OF MAJOR EVENTS — SAN DIEGO WATER STORY

1815	Old Mission Dam and aqueduct completed (approx.)
1834	Missions secularized
1850	San Diego City and County formed by state
1864	Severe California drought commences
1884	Great San Diego Flood
1886-1923	San Diego's (private capital) dam building spree
1901	Municipal ownership of San Diego water system
1905	Owens Valley project commenced by Los Angeles
1905	Imperial Valley flooded and Salton Sea created
1909	Imperial County formed out of San Diego County
1911	Imperial Irrigation District (IID) formed
1913	Los Angeles Aqueduct from Owens Valley completed
1916	Hatfield Flood
1928	Metropolitan Water District (MWD) formed by state statute
1931	242-mile-long Colorado River Aqueduct (CRA) begun by MWD; terminal storage Lake Matthews in southwest Riverside County
1935	El Capitan Dam completed by City of San Diego
1935	Hoover Dam completed by U.S. Bureau of Reclamation
1938	Parker Dam completed by MWD
1940	All-American Canal completed by U.S. Bureau of Reclamation and IID
1941	Colorado River Aqueduct (CRA) completed by MWD
1943	San Vicente Reservoir completed — terminal storage for future 1st and 2nd San Diego Aqueducts
1944	San Diego County Water Authority (SDCWA) formed by state
1946	SDCWA annexed by MWD
1947	San Diego receives Colorado River water via 1st San Diego Aqueduct, built by U.S. Bureau of Reclamation

1954	2nd San Diego Aqueduct completed via new Lake Skinner in Riverside County to San Vicente Reservoir in San Diego County
1960	3rd San Diego pipeline of 2nd San Diego Aqueduct complete; terminal storage Lake Miramar with extension south to Lower Otay Reservoir
1960	State Water Project (SWP) approved by voters
1968	Oroville Dam complete on Feather River
1972	Completion of California Aqueduct to Los Angeles
1972	4th San Diego Aqueduct pipeline complete; terminal storage Lake Miramar; treatment plant (also treatment plant at Lake Murray)
1978	Completion of California Aqueduct to Lake Perris
1978	San Diego first receives Feather River water via SWP, blended with CRA and treated at Lake Skinner
1982	Peripheral Canal for Sacramento River-San Francisco Bay delta defeated by voters
1982	5th San Diego Aqueduct pipeline complete; terminal storage (near San Marcos)
1987-1991	Severe southern California drought
1988	San Diego-IID water transfer agreement announced
2001	Diamond Valley Lake, between Lake Perris and Lake Skinner in southwest Riverside County, completed off-line
2003	San Diego–IID water transfer agreement finalized
2003	Olivenhain Reservoir completed off-line by SDCWA as part of the Emergency Storage Project
2003	Quantification Settlement Agreement (QSA) signed by local, state, and federal agencies
2004	San Diego receives water via Imperial Valley agreement

CHAPTER 1
THE ESSENTIAL BACKGROUND

THE LAND WHERE WE LIVE

The words "natural water" invoke myriad images ranging from shimmering lakes to turbulent seas, from gurgling streams to raging rivers, from bone-dry waterholes to inundating floods. All these outward manifestations plus the inner, life-giving quality of water are well known to the arid San Diego region. Landscape colors seem to reflect these images: a dramatic combination of the dull brown of the desert with the bright green of lush vegetation and the vivid blue of the ocean. Certainly not a match for the varicolored spectrum of the rainbow, but nonetheless a brown-green-blue triad that symbolizes a transition from desert stillness to water-made life.

This account begins with that arid landscape as the Spanish explorers and settlers began to re-learn the lesson they should have remembered from their years in conquering Mexico — that those who occupy this land are excruciatingly dependent on an adequate supply of precious water. The explorers Juan Rodríguez Cabrillo and Sebastián Vizcaíno came to San Diego Harbor in 1542 and 1602 and then left quickly as explorers historically and customarily did.

What is most significant to this narrative is not the explorers or the natives before them, but the nature of the land that the Indians lived on and the explorers discovered. The Kumeyaay Indians could have told Vizcaíno that the river that beguiled him with its "sweet and good water" was not as reliable as it may have seemed. He may have asked about rainfall and the Indians undoubtedly nodded agreeably, having neither the ability nor the desire to communicate much to these strange people whom they rightfully and instinctively disliked and mistrusted.

None of their homeland or New Spain experiences prepared the Spanish fully for this land's wild climactic and topographic contrasts. From a distance the landscape above the harbor may have appeared rather benign to those early explorers, but up close the hills often gave way to boulder-strewn mountainsides where cactus was more plentiful than even chaparral, let alone trees. The northeastern mountains had forests of coniferous trees but to the southeast (behind the harbor), only piles of rocks dominated the landscape.

1

The rain fell in the mountains and valleys and there were sporadic springs, but the soil was mainly infertile. At lower elevations, the terrain seemed more welcoming but was almost invariably too dry. The much-needed rain deserved to be hailed as "Godwater," as it often was by Native Americans. Walter Prescott Webb, writing for *Harper's Weekly,* called the West of the eighteenth century "a semi desert with a desert heart." That may well have been too extreme for San Diego County, but climatologists have classified it as "near desert" and when the Godwater did not fall, it would be arid indeed.

Nature and geopolitical circumstances have isolated San Diego County into a corner of the nation bounded by a desert, a mountain range, and an ocean. At one time the county extended all the way to the Colorado River. Now it is smaller, with 4207 square miles of hills, valleys, and mountains — and yes, beaches. The county forms a rough rectangle (about 60 miles NS and 70 miles EW). Beautiful beaches line the coast and the climate is akin to that of the lands bordering the Mediterranean Sea.

From high Cuyamaca Peak (6515 feet), summers 10 miles to the east are among the hottest, and 35 miles to the west are among the coolest, known in the United States. These mountains, where the rainfall is most bountiful, provide the watershed areas for six rivers that flow through the county, generally east to west. In those mountains are natural springs and streams that bring year-round greenery to some of the valleys. Beyond those mountains lies Imperial Valley, once in centuries past a part of the Gulf of California when that gulf extended as far north as the San Jacinto Mountains. For many years, Imperial Valley was part of San Diego County.

Rainfall may reach 40 inches in the mountains but the coastal plain, about 900,000 acres where by far most of the people (about 3 million in the year 2003) live, often gets along with less than 10 inches, and sometimes less than half of that. Rainfall runoff has never been a reliable annual source of water; the San Diego area has one of the greatest differences in rainfall runoff between the wettest and driest years of any region in the United States.

THE WATER GODS SMILED, THEN FROWNED

Except for an early drought, rain was most often plentiful during the early years of the 1800s. During the years that the missions were operative, the Spaniards often overestimated their power to subjugate a superficially attractive but basically inhospitable land and to truly "civilize" the often-rebellious Indians. And they wholly failed to realize the tremendous difficulties that would arise from a drastic lack of water, a commodity essential to colonization.

The disillusionment began when Don Gaspar de Portolá, Father Junípero Serra's expedition leader, set out to find the "sweet and good water" that his predecessor explorer Vizcaíno had reported. Exploring north and east from the

San Diego Harbor, Portolá learned the hard way that marches were difficult because, as he wrote in his journal quite simply at the end of the first day, "There was no water for man or beast."

Serra and his companions soon learned the same water lesson the hard way when they tried to plant crops at the newborn Mission San Diego de Alcalá. First, they discovered that the promising San Diego River often became an "upside down river" because, as a native explained, it runs on the surface in the winter but in the summer, it runs "upside down" under the sand. In addition to that problem, crop failures resulted from the river's water vagaries — first too much water flooded the planting and then too little water parched the planting. The padres soon sought a new mission site. Upstream 6 miles they went to find better water, better land, and more peaceful Indians. After application to the Spanish Viceroy, Serra was given the go-ahead for the mission relocation but with an historic caveat governing the San Diego River:

> "His Grace," wrote the viceroy's second secretary, Julio Ramon Mendoza, to Fr. Serra in 1773, "reposes herein a sacred trust to acquire and administer the waters of aforesaid stream for the common benefit of all the people, both Gentiles and converts, who reside, or in the future may reside within the jurisdiction of Mission San Diego de Alcalá."

The "Gentiles" were, of course, the native Indians. And the words "common benefit" were shorthand for the Spanish rules already developed that provided for the utilization of scarce water for all the people who needed it.

Upstream several miles north of the mission, Indians showed the padres an exposed granite shelf that could be used as the base for building a crudely cemented rock dam when the river flow was reduced to a trickle in the dry season. Old Mission Dam took months to build and then years to refine for water distribution. It was 244 feet long, with a 12-foot spillway and gate, about 12 feet high, and 10 feet thick. Larger stones were piled on top of each other and held together by mortar to form the sides of the dam; smaller rocks covered its interior. Then, after years of work by the Indians, the flume was built to carry the water over 3 miles down the rocky, steep-sided Mission Gorge, and then through a tile-bottom ditch for several more miles to the mission. This was a remarkable accomplishment for non-engineers working with primitive materials and methods. Remains of the old dam still stand today.

The agricultural growth that subsequently began was fortunately accompanied by years of mostly moderate rainfall. There were some drought years, but overall the streams, springs, rivers, and above all the rainfall were sufficient to salve the peoples' thirst, provide an adequate water supply, and constantly renew the grass that fed growing herds of cattle and sheep. Mission San Diego de Alcalá became a major operation after several decades, with thousands of acres tilled, 467 olive trees and 8600 vines planted at the mission, plus another 8000

Old Mission Dam   *Bill White*

vines planted in El Cajon Valley. For livestock, it ranged 4500 cattle, 13,250 sheep, and 200 horses. Making the entire economic system possible was the steady seasonal rainfall which kept the green grass growing for the cattle and the irrigation ditches flowing for the crops, orchards, and vineyards.

During the second decade of the 1800s, Mexican yearning for independence had become a driving force, culminating in victory over Spain in 1821. The missions were secularized in 1834. Then, the new Mexican government's practice of making extensive land grants led to the growth of the large ranchos that soon dominated California.

These huge land grants went to the elite, men with a Spanish heritage who carried the title "Don" and called themselves the *gente de razón* (men of reason). What has been called the "Days of the Silver Dons" ensued as their herds grew and their ranchos took over the immensely profitable hide and tallow trade which was centered at La Playa on Point Loma. For decades, from 1820 through the 1850s, from Monterey south for over 400 miles to (and including) present-day San Diego County, the valleys and hills became a vast, unfenced grazing ground. The rainfall was plentiful in those years and every rancho in the San Diego area was located near rivers, springs, and streams sufficient for both people and livestock.

The years dominated by the Silver Dons brought a dramatic change in water rights throughout California. Spanish law had emphasized that water should be held for the common benefit. No person had a right to a specific volume of water, and when disputes arose the water was to be allocated "with equity and justice to all." Fairness controlled and no paramount or exclusive rights could be acquired. These concepts were embodied in what was called the "Plan of Pitic," developed by Spanish overlords for one pueblo but subsequently applied to others throughout the Southwest. Then, as the pueblos grew, they increasingly stressed their claims to the water in nearby rivers and streams. It was argued that the 1848 Treaty of Guadalupe Hidalgo which ended the Mexican War protected the pueblos' water claims since it contained a commitment to continue existing rights.

Then came the gold miners and rancheros. They diverted water from adjoining rivers and streams to conduct their private enterprises and claimed what were called "riparian rights" over river water. The law developed that an appropriator of river water could possess a right superior to a riparian user if the appropriator had begun using water before the riparian taker came along. Time of taking became the test. "First in time, first in right" became the battle cry of those who got to the river first. Ownership of wells and springs was never disputed but rights of appropriators and riparian rights ruled on the rivers. That "law of the river," as it was called, controlled water right disputes throughout the state until it ran head-on into the "pueblo rights" asserted by the cities.

California had never seen a drought like the one that followed quickly on the heels of the "great flood of 1862," dealing the final blow to the reign of the Silver Dons. Their mainstay was cattle and their herds had grown to over 1 million head by 1860, grazing constantly on the hillsides of California. As the grass shriveled into dust, cattle by the thousands starved and literally dropped in their tracks. Fences were built with the bones from dead cattle and, as the *Los Angeles News* noted, "the largest rancheros kept their men busily engaged in skinning the hides and thousands of carcasses strew the plains in all directions." Waterholes used by the cattle disappeared as riverbeds dried into cracked earth. In vain attempts to save grass for the cattle, huge herds of horses were slaughtered or driven over ocean cliffs and the brinks of canyons.

Then, to make matters worse, a smallpox epidemic swept the state. Thousands suffered from the oppressive drought and the fatal epidemic. At the end of 1864, John Forster, owner of the huge Santa Margarita Ranch in northern San Diego County, stated that "the whole country from north to south became almost depopulated of cattle." In sad words that have echoed down the years, Forster concluded, "there was a perfect devastation; such a thing was never before known in California."

HUMAN FLOODS AND WATER FLOODS

When California became the thirty-first state in 1850 and San Diego was incorporated as a city, there were only 650 residents. "Dreary and dismal" was a frequent travelers' verdict. Ships did occasionally call for the hide and tallow trade and one logged the sailors' description of San Diego as "quiet, dull, desolate, hot and going to ruins." The log of another went even further, terming it "the meanest place of all, not a morsel of fresh provisions and the town is four miles from anchorage."

There were two water sources for the communities of Old Town and La Playa. La Playa, later becoming Roseville on Point Loma, used water from a well on North Island; barrels were carried across the channel on a rowboat. For Old Town's supply, watermen sank large barrels into the sand at the San Diego River to collect water; they then stored it in cisterns. Filling their wheel-mounted barrels at the cistern, young boys roamed the streets, hawking their precious fluid for 25 cents a bucket.

Water boy   State of California Water Atlas

San Diego came to life after the end of the Civil War. Many thousands of people followed Horace Greeley's famous advice, "Go west, young man." Some of the new residents captured rainwater from their roofs, storing it in private cisterns. Wells were dug by others, and those with enough money built windmills to bring the water to the surface. Finally, the city sank a well in 1873 but residents turned their noses up at the quality of this water, using it for bathing but not drinking. Nor for laundry; one washerwoman said, "When you put soap in it and tried to mix it, it turned into a kind of chalk." Wells were eventually drilled in Mission Valley and water was pumped up to storage tanks and thence to the city dwellings. But the quality of this well water was not the best. A standard joke of that time was, "We boiled it, we screened it, we boiled it again and then we drank something else." Better water was finally found in upstream, deeper wells and it was pumped to a reservoir in University Heights (near present-day Texas Street). When that proved too expensive, a tunnel was built that carried the water to a new reservoir at Fifth and Hawthorne streets.

In the 1870s and 1880s, as Joaquin Miller, known as the "Poet of the Sierras," wrote, "San Diego was as suddenly born as if shot from a gun." Cheap land, cheap railroad fares, and bombastic promoters turned what had been a trickle of new settlers into an almost daily human tide that inundated the city ("San Diego's Golden Era," it came to be called). Population doubled and then redoubled. The growth was also dramatic in what was called the "backcountry." Thousands of land-hungry people from the East and Midwest succumbed to the persuasive pitches of real estate salesmen, buying lots all over San Diego County. People bought land they had never seen, land that might not even exist. When the land sales finally tapered off at the end of the 1880s, San Diego's population had risen to about 35,000, quadruple what it was in 1870.

Meanwhile, the San Diego pattern continued of too much and then too little water. Flood and drought did not have regularly recurring cycles, but newcomers soon learned that, sooner or later, one was sure to follow the other. The fury of the floods was mostly felt in the settlements. These had been located, quite naturally, near the rivers ranging from the Santa Margarita and San Luis Rey rivers on the north down to the San Dieguito River, then the San Diego River, Otay River, and Tijuana River. Any extremely heavy rainfall would bring a wall of water down each of those riverbeds, and nothing could stand in its way.

One of the earliest recorded series of heavy floods in California commenced in December 1861 and continued through the following summer in many areas of the state. The San Diego River rampaged after 15.75 inches of rain fell during the winter of 1861-62. Fast-flowing water from the unusually heavy downpour carved gullies and canyons into previously rounded hills and the San Diego River cut a new channel into San Diego Harbor.

One of the worst floods in San Diego's history occurred in 1884 following 25.97 inches of rain, the greatest amount ever recorded. But 1884 paled alongside 1916 when the "Hatfield Flood" came on the heels of a ten-year drought. Charles Hatfield was a traveling rainmaker who boastfully offered to solve San Diego's water problem. "I will fill the Morena Reservoir to overflowing for the sum of ten thousand dollars," he promised. The city council voted unanimously to accept the offer. Up near Lake Morena went Hatfield's crude wooden tower with a platform on top and the vapors from his chemicals were soon wafting to the sky (some said they smelled like limburger cheese). Amazingly, it began to rain the very next day, a steady rain that wouldn't quit. Three days later, the rain reached torrential proportions in many parts of the county.

Reports streamed in of backcountry bridges being washed out, hundreds of cattle drowning, roads made impassable. The Santa Fe train was stalled near Oceanside for forty hours; the railroad bridge at Del Mar subsequently collapsed into the roaring water. No trains ran for thirty-two days. The coast highway bridge met the same fate and all the rivers overflowed their banks. Mission Valley became a water thoroughfare as the San Diego River covered the valley with a mile-wide raging flood.

Water levels at Morena Reservoir climbed rapidly. Upper Otay Reservoir filled, and when the spillway of the Upper Otay Dam proved tragically inadequate, water roaring over the top of the dam soon washed away supporting fill at its base. The dam's steel core, exposed to the water, opened like a gate and a veritable wall of water roared down the canyon. As one observer said, "All hell broke loose." Uncounted millions of gallons of water went on a seven-mile rampage to the bay, sweeping everything before it, including houses, ranch outbuildings, citrus groves, and chicken farms. Layers of topsoil were stripped from the land which was left like a gravel bed; farming could not be resumed. Near the mouth of the valley, the huge water wave hit a 3000-foot-long railroad embankment with a bridge; both were immediately swept away.

But Hatfield had fulfilled his contract; Morena was full. Fearful of being lynched by angry farmers, Hatfield "got out of Dodge" as the saying goes, leaving town during the night. He never received his $10,000. The City simply welched, pointing out that Hatfield had no written contract and attributing the unheard of rainfall to "an act of God," to the delight of the insurance companies.

In the years and decades to follow, more floods came but none wreaked the devastation caused by the memorable "Hatfield Flood" as it came to be called. Those watching saw hundreds of thousands of acre-feet of water flow out to sea — enough to see San Diego through many decades of dry years. But that water disappeared almost instantly and in a frustrating, seemingly inevitable, never-ending cycle, San Diego became known as an area where floods always followed droughts in devastating succession.

Just as people in San Diego have over the years achieved a truce with "sunshine dollars" by accepting lower pay to enjoy the climate, so they have learned to live with the twin disasters of drought and flood. The reasoning is: "Sure, the dry years on occasion parch the hills and valleys and the fury of floods is felt from time to time in the towns and cities but overall the climate makes it all worthwhile." Floods and drought, like tornadoes in the Midwest, hurricanes in the Southeast, and blizzards in the East, come only "every now and then," so they can be tolerated.

Sweetwater Dam failure, 1916
San Diego Historical Society Photograph Collection

CHAPTER 2
DAM BUILDERS AND WATER HUSTLERS

"As any small boy knows, the presence of running water is a reason to construct a dam. Water hustlers began when man moved to the desert and devised a manner of moving water from where it was to where it had never been before. Water hustling became the 'second oldest profession in the world.'" Carl Joseph Courtemanche, *Utilization of Water in San Diego County from 1890 to 1940* (SDSU master's degree thesis on file at San Diego Historical Society archives).

YEARS OF THE PRIVATE WATER COMPANIES

Providing water for people to drink and agriculture to thrive was for years in San Diego the responsibility of private individuals, starting with those who found their own water in the countryside and the hardy men and boys who collected water and sold it by buckets door-to-door in the city. They were succeeded in the late 1800s by private companies, first the San Diego Water Company, which drilled deep wells in Mission Valley and constructed concrete reservoirs to store the water and pipelines to carry it to homes.

Soon, though, there came a different breed, people who "read the tea leaves" and concluded that much more water would be needed. Not much imagination was required to foresee a huge demand for water in an arid county with great agricultural potential and a growing urban population. Why dig wells and build concrete tanks, these men thought, when rivers could be dammed to create huge reservoirs?

SAN DIEGO'S SPECTACULAR DAM BUILDING SPREE

A dam building spree ensued in San Diego County late in the 1800s that has never been matched anywhere else in the nation. By the end of that century, "San Diego County could be accurately described as one of the major focal points of dam construction in the world," wrote Philip R. Pryde in *San Diego: An Introduction to the Region* (1995). Someone once described the construction of a dam as man's ultimate way of thumbing his nose at God. San Diego's dam builders could have replied that they were merely giving God a helping hand and that it is not a sin to make money while doing so.

But sites where these dams could be built were fewer than the entrepreneurs originally thought. Unlike those elsewhere in the nation, San Diego County's rivers run a very short course from their beginnings to the ocean, and most drop down fairly precipitous mountain slopes until they come very near the coast. In this limited water run, there are few locations for huge dams requiring a deep valley with a narrow spot where they can be anchored to bedrock at the base and fastened structurally to rocky walls on each side.

First to be built was the Cuyamaca Dam and Reservoir (Lake Cuyamaca). The original plan for a masonry dam was scrapped and a simple earthwork dam was built with good, cheap clay found nearby. Mule teams pulling scrapers moved the earth into piles where it was loaded into wagons that more mule teams pulled to the dam site. As the dam was formed, it was covered with 8 inches of stone riprap. Two spillways were added at the top, along with wooden towers to hold iron gates sliding in wooden grooves for controlling the water outlet. Completed early in 1887, the new dam that foretold the future for San Diego's water supply was 41.5 feet high; the rainy season soon filled the reservoir.

The San Diego Flume, a miles-long wooden flume constructed to move the water down the mountain came next. This "silly scheme" to bring water from the top of the mountains down to the city was derided by some observers. The *San Diego Union* reported that it was discussed by promoters "at the risk of provoking a legal inquiry as to their sanity." However, after it was built by San Diego Flume Company, the *Union* quickly reversed itself, concluding editorially and grandly that "this great flume will solve San Diego's water problems for all time."

More than 800 horses in teams of 6, 8, or 10 pulled 100 wagons in caravans of 5 or 6, carrying about 9 million board feet of redwood lumber to build the flume and supporting structures. Right-of-way negotiations involved hard bargaining and usually water, not money, was offered as compensation. Initially, it was difficult to get skeptical landowners to even negotiate since no one was taking the project seriously, but that changed with time. Very gradually the necessary 50-foot-wide right-of-way was assembled. The length of the flume would be 35 miles. Designed to be built with redwood planks on a pinewood base, it would be 70 inches wide inside and 14 inches deep. The water depth would be 10 inches.

The necessary tunnels (eight in all) and wooden trestles (a total of 315) were laboriously completed by men and mules working long hours. Hundreds of Chinese workers, "coolies" they were called, were brought down from San Francisco to work on the project. The longest trestle was Los Coches at 1794 feet. Needed to cross one of the deepest passes was the highest trestle at 82 feet. These were aptly called "skeleton trestles," built in sections assembled flat on the ground and then raised by mule teams pulling in harness on ropes through pulleys to become part of the structure.

San Diego Flume, Sweetwater Trestle
San Diego Historical Society Photograph Collection

The San Diego Flume ran down the south side of the river to the El Monte Tunnel near Lakeside. The route crossed the El Capitan Grande Indian Reservation; the Indians were paid $100 per mile of flume plus a guarantee of all the water they would need. After the El Monte Tunnel, the water was turned south to follow around El Cajon Valley to its terminus near La Mesa, where it was fed into a steel pipeline that carried the water to the city mains. Along the way, the flume supplied water through local water districts to the irrigation projects of farmers and orchard growers.

On opening day, boatloads of celebrities took a noisy, frolicking ride all the way down the San Diego Flume. In the city, San Diego staged a welcoming celebration on Washington's Birthday, February 22, 1889. A massive parade with numerous floats dazzled bystanders on Market Street. Nozzles on street corners sprayed fountains over one hundred feet into the air. People rushed to taste the water, yelling, "Oh, it was delicious!" and exclaiming over the "pure mountain water." Actually, it wasn't. The Flume Company had to borrow from San Diego Water Company enough water to fill the pipes; the "mountain water" did not come into the system until the next day. Thereafter, the San Diego Flume Company and the San Diego Water Company together supplied the city's water needs. And all of it was San Diego River water.

After Cuyamaca Dam came the dam on the Sweetwater River built by the Kimball brothers about ten miles east of the city. Sweetwater Dam was the first of the more modern dams using a design called the "thin single gravity arch" which relies on the weight of the concrete to hold it in place while the arch construction transfers the water force onto a solid rock abutment on each side of the dam. Sweetwater was the highest structure of this type when it was completed in 1888. Constructed of stone heavily impregnated with iron so that it weighed 200 pounds per cubic foot, the dam was designed to be the highest in the United States. Stone for the dam was quarried locally, just a mile downstream; the cement used to marry the stones was shipped from Belgium.

One of the most prominent dam builders was the Southern California Mountain Water Company, a long name usually shortened to Mountain Water Company, organized to supply water for irrigation in the South County area, primarily for lemon orchards. The company was formed in 1895 by Elisha Babcock, a powerful San Diego businessman, and John D. Spreckels, the silver-spoon son of Claus Spreckels, the multimillionaire sugar king who had become a San Francisco business magnate. Babcock had built the Coronado Water Company and was an early shareholder-investor in the San Diego Flume Company. Spreckels was a relative newcomer to San Diego whose announced goal was to "make big money" from land and water development. When the Babcock-Spreckels South County plan was announced, it was hailed in the *San Diego Union* as the greatest irrigation project in the nation.

Mountain Water Company's overall plan commenced with the Morena Dam and Reservoir to be built in the headwaters of the Tijuana River. Water released from Morena would flow by gravity down to the planned Barrett Reservoir, a diversion reservoir needed to provide a constant flow of water into the Dulzura Conduit that would carry the water 12 miles to Dulzura Creek, down which it

Morena Dam   *San Diego County Water Authority*

would flow another 12 miles into the two Otay Lakes, Upper and Lower. A daring project, the concrete-trench Dulzura Conduit was described by an anonymous observer as "a tortuous triumph like a big water snake winding across the mountains, clinging to breathlessly sheer cliff sides, rounding embarrassing shoulders, crossing great chasms on long, slim trestles, gliding through tunnel after tunnel along its perilous journey."

Along the way, water could be sold for both domestic and agricultural use. The end would be the Chollas Reservoir just a few miles east of downtown San Diego, where a filtering plant could ready the water for the hoped-for sale to the City of San Diego. This would be accomplished by piping the water to the City's University Heights Reservoir and thence into the city distribution system.

Morena Dam, commenced in 1895 and reaching a towering 150 feet when completed, rests on bedrock of solid granite. The dam, arch-formed and having an inward vertical face at a 45-degree angle, was built with huge blocks of granite, each weighing about 4 to 8 tons, that were quarried from the sides of the gorge. These blocks were set in cement, surfaced with concrete and anchored to the bottom through parallel vertical slits. Bracing the dam were thousands of tons of broken rock dropped behind the dam into wooden forms faced with concrete slabs from huge buckets crossing the gorge on a high-hung trolley line. Sixteen feet thick at the top, Morena was called the biggest dam in

Barrett Dam   *San Diego County Water Authority*

America and possibly the world by *Harper's Weekly* in January, 1898. Morena Dam cost the entrepreneurs who built it about $5 million.

Barrett Dam, built to provide the diversionary flow of water it received from Morena Reservoir (Lake Morena) into the Mountain Water Company system, was a massive concrete dam having the upstream face curved on a 400-foot radius to form a single arch across the canyon. While it was being completed, flooding from heavy rains commenced and the lake behind the dam rose unexpectedly fast as the rains continued. A closely watched race began to keep the dam going up faster than the water.

Last to be built in the Mountain Water Company system were the two Otay dams, Lower Otay first and then Upper Otay. For Lower Otay, Elisha Babcock designed a new construction method using a series of steel plates one-third inch thick, riveted together at the center of the dam to form a large wall which was anchored at the bottom with a massive concrete pour. The dam was locked on each side into solid rock with abutments. The steel plates, formed during construction, created a pyramid that was used as a crane to move concrete and rock obtained from a nearby quarry. Explosives were used to break the rock into small pieces; 100,000 pounds loosened 250,000 tons of rock. The steel core was shielded on each side by a foot of concrete and then, using wooden forms, huge loads of random-sized rock were dumped against both sides. A spillway on top and an outlet tunnel completed the Lower Otay Dam; it was 134 feet high and 565 feet wide and held 42,000 acre-feet of water.

The Upper Otay Dam was a thin-arch concrete dam like Sweetwater and it had one construction fault. Observers saw a small stream of water flowing from the base of the dam after it was built. This was intentional because downstream farmers had complained that the dam cut off their water supply in violation of their riparian rights, and they threatened to dynamite the dam. To placate them, the construction company placed three large pipes underneath the dam and filled the area around the pipes with porous cement which allowed enough water to seep through to satisfy the farmers' needs. This construction decision had tragic aftereffects. It created a "soft spot" in the dam which contributed to its failure during the Hatfield Flood described earlier.

Farther north, William Henshaw, working with land and water developer Ed Fletcher, formed the San Dieguito Mutual Water Company to concentrate on developing the San Dieguito and San Luis Rey rivers to serve North County agricultural needs. Their company also built the Hodges Dam, named after a Santa Fe Railroad executive. Water from Lake Hodges went to the frost-free coastal area ranging from Del Mar north to Carlsbad, which rapidly developed huge vegetable-growing operations and provided the location for the first major planting of imported tropical avocados.

That railroad executive named Hodges originated the scheme to import from Australia thousands of eucalyptus trees for planting at the 4000-acre Rancho Santa Fe Spanish land grant also owned by the Santa Fe Railroad. Hodges thought that the trees would make marvelous railroad ties. When the wood proved to be too soft, the Santa Fe lost millions of dollars and Hodges lost his job. Years later, the land was sold to Bing Crosby and developed into the charming North County village named Rancho Santa Fe. The village residents owe a debt of gratitude to Hodges for their trees and also to Fletcher, the village's first manager, who built many of the hillside roads that still serve the village.

The Hodges Dam created a stir in San Diego engineering circles because it was built with a new and dramatically different hollow, multiple arch design which cost much less than the single-arch gravity dams built previously. Hiram N. Savage, who had designed the Barrett Dam and became the City's water engineer, publicly proclaimed Hodges Dam "unsafe" and, according to Fletcher's *Memoirs*, "went out of his way to ridicule the dam." Savage was equally critical of the multiple arch, 117-foot-high Murray Dam also built by Fletcher with financing from his partner, James Murray, after whom the dam was named.

Both the Hodges and Murray dams were designed by John S. Eastwood who found a booster and patron in Fletcher when he was satisfied that multiple-arch dams could be built with far less money. Throughout the west, Eastwood for years battled John R. Freeman, defender of the conservative dam designers like Savage, who favored the "massive, solid" single-arch dams they

Hodges Dam   *City of San Diego Water Department*

Murray Dam   *San Diego County Water Authority*

thought were desirable to engender public confidence; no matter that they were vastly more expensive. These men dismissed Eastwood's designs as "airy arches with lace curtain effects." Years after Hodges Dam was built, the conventionally minded California authorities forced the building of huge concrete supports on the downstream face of the dam, supports that Eastwood would have dismissed as being a waste of money. The controversial supports are clearly visible to drivers on Del Dios Highway from Escondido to Del Mar. For the whole story on John Eastwood's controversial career, see Donald C. Jackson's *Building the Ultimate Dam* (1995; University of Kansas Press).

Ed Fletcher recorded in his *Memoirs* that he went to great pains to get insurance protection from Lloyds of London for both the Hodges and Murray dams and was pleased to brag publicly that his premium cost was substantially less than that being paid by the City for insurance on the single arch dams built by Savage. Of course, as people in San Diego now know, Savage's dire prediction about the multiple arch dams was faulty. The Hodges and Murray dams have easily survived the years. Lake Hodges, originally a water source for the agricultural area from Del Mar to Laguna, is now primarily a storage reservoir.

FINALLY, ALL SAN DIEGO COUNTY RIVERS WERE DAMMED

John Eastwood's mantra was, "The California slogan e're should be that 'tis a crime to let our rivers reach the sea." And he was proud to teach how this could be done with private capital. Remarkably, all the above-described San Diego County dams and several others were financed entirely by individuals and privately owned companies (until Barrett Dam was completed by the City). Uniquely, no government bonds, no federal, no state, no city money was asked for or used and this feat has not been duplicated anywhere in the nation.

When the massive construction projects were finished, all six of the east-west running rivers in San Diego County had dams and reservoirs in their watershed areas. Going north to south, the San Luis Rey River had Henshaw; the San Dieguito had Hodges; the San Diego River had Cuyamaca; the Sweetwater River had Sweetwater; the Otay River had Upper and Lower Otay; Cottonwood Creek (which runs into the Tijuana River) had Morena and Barrett. According to historian Austin Adams in *The Story of Water in San Diego*, these reservoirs "removed the element of water from the anxieties of all the future." Little did Mr. Adams know.

The dam builders and men called "water hustlers," who sold the water made possible by the dams, shaped the development of San Diego County's water supply sources. Other entrepreneurs formed a variety of water development companies over the decades at the end of the 1800s and the beginning of the 1900s. Many of the reservoirs in San Diego County today share a heritage with one or more of those companies. Several of those companies failed and

William E. Smyth, in his *History of San Diego* (1907), describes the water hustlers' activities as constituting "a prolific source of controversy marked by acrimonious discussions and sharp divisions in the community."

The major survivors of these water development controversies and financial maneuverings were Spreckels-Babcock's Mountain Water Company, Fletcher's Cuyamaca Water Company, Kimball brothers' Sweetwater Company, and the North County systems of Henshaw and Fletcher. And of these four, clearly the two most powerful were the Cuyamaca and Mountain water companies headed by two imaginative, hard-driving men who vied with each other for years in a series of conflicts that involved major development projects, competing dam sites, huge financial stakes, power-hungry politicians, and greedy investors.

THE SPRECKELS-FLETCHER DUEL

John D. Spreckels and Edward Fletcher were two antagonistic protagonists in a constant, decades-long conflict that could, with equal justice and depending on viewpoint, be called either two businessmen competing for pockets full of dollars or one businessman competing with a visionary who put his own business interests in second place.

Spreckels was recognized for years as the "number one" citizen of San Diego. The newspapers he owned, including the *San Diego Union*, were criticized as publicity sheets for the San Diego business establishment. Spreckels was made from the mold that has produced many successful entrepreneurs in America. They often made it their "business" to become a moving force in the affairs of the city where they lived and worked, using their power to help the community grow and also generously to help themselves. Spreckels clearly wanted San Diego to become a first class metropolis but he also subscribed to the "What is good for General Motors is good for the United States" mind-set. He believed strongly that what was good for John D. Spreckels was good for San Diego.

Spreckels' frequent opponent was Edward Fletcher, who at age 16 arrived in San Diego in 1888 after a tourist-fare trip from Massachusetts with $6.10 in his pocket and began what could justifiably be called a "Horatio Alger" story. Starting as a small wholesale produce trader, Fletcher grew his business by buying eggs and butter cheaply in Wisconsin, storing them on ice during the winter, and then selling the products at high summertime prices. For the rest of his life, Fletcher concentrated on land and water development in north and east San Diego County. He also served as a state senator for two terms and after being called up as a member of the National Guard to serve in the San Francisco earthquake crisis in 1906, was promoted to the rank of lieutenant colonel. From then on he was known affectionately to his friends as "Colonel Ed."

There is no surviving personal description by Spreckels of Fletcher but the newspaper he owned, the *San Diego Union*, charged Fletcher with bribing a city councilman, often accused him and his cohorts of being involved in "wild and visionary schemes," and criticized their development proposals as "creations of their imaginations." A local newspaper once called a Fletcher project: "FFFF: Freaky Fletcher's Fancy Flight." On the other hand, Fletcher in his *Memoirs* described his duel with Spreckels as "a conflict with big financial interests, greed, graft and jealousy," and charged Spreckels with owning "nearly everything of value" in the city of San Diego which became virtually a "one man town for years."

"One man town" was surely an exaggeration but Spreckels did own not only the powerful Mountain Water Company and the Hotel Del Coronado, most of Coronado, and the Silver Strand, but also developed Mission Beach and its amusement center and built the *San Diego Union* building, the Spreckels Theater, the San Diego Hotel, and the Bank of America building. Spreckels' short-line railroads served much of San Diego County, and his San Diego & Arizona Eastern Railroad built a branch line that ran daily from San Diego to stations in downtown Tijuana and at the newly constructed Agua Caliente Racetrack. These sites attracted gamblers from throughout the West to the point where the *Los Angeles Morning Tribune* complained about the "racetrack gambling hell at Tijuana."

Fletcher and Spreckels met head-on when Mountain Water Company tried in 1897 to take over the contract held by Fletcher's Flume Company to supply water to the City of San Diego. Competing bids were submitted and the city council had to make the decision. Today's often-condemned "negative campaigning" pales in comparison with the personal assaults leveled during the public fight that ensued during the mayoralty contest where the water contract was an issue. Fletcher's and the Flume Company's candidate for mayor was attacked in the pages of the *Union* as: "charlatan, faker, traitor to the people, a cigar-and-smile trickster, political quack, pretender, clown, cheap gambler, bamboozler and liar." In the paper favorable to Fletcher, Spreckels was described as a "millionaire trust monopolist who was a cloven-hoof and fork-tail devil who roasted his victims over his private grill to gratify his monopolistic appetite."

The Flume Company's bids were eventually accepted, but the company (with Fletcher as owner and general manager) could not satisfy the city's water needs. After a few years, Spreckels' Mountain Water Company became the water supplier to the city. The Flume Company continued to operate in the El Cajon and La Mesa areas but water leaks constantly kept the flume in operational trouble until Fletcher solved the problem by coating the entire surface of the flume with rubberoid roofing material (cost: $44,000), cutting the leakage down to 10%. Fletcher merged the Flume Company into his Cuyamaca Water

Company. By then, the flume had distribution lines running to Lakeside, El Cajon, La Mesa, Lemon Grove, Spring Valley, and East San Diego (not a part of the City of San Diego at that time).

In 1913, Fletcher offered his Cuyamaca Water Company system to the City of San Diego for $300,000. With Spreckels openly opposing acceptance of the offer, the city turned it down. In 1916, Fletcher renewed the offer, raising the price to $745,000. Again, with Spreckels leading the opposition, the offer was rejected, much to Fletcher's unhappiness. In the early 1920s, Fletcher negotiated for a merger with the Sweetwater system and made another offer to sell the Cuyamaca system to the City of San Diego, this time for $1.2 million. Both of these deals fell through, again with Spreckels aiding and abetting the opposition.

Meanwhile, in 1901 San Diego had formed the Consolidated Water Company and brought under that municipal umbrella the old privately owned San Diego Water Company together with Mountain Water Company's delivery system within the city. However, the City continued to buy water from the Flume Company and, later, from the Mountain Water Company's Morena-to-Otay-to-Chollas system when Spreckels managed to outmaneuver the Flume Company and take over the City water business. Then, in 1912 and 1913, two bond issues raised the money for the City to buy the entire Morena system. Spreckels complained that he lost money on the sale, but needed the money to finance his pet project, the San Diego & Arizona Eastern Railroad.

Fletcher made his last offer to the city in 1923: $1.4 million for the Cuyamaca Water Company system. The City turned it down with Spreckels arguing against the offer. Spreckels in 1923 poured out his pent-up feelings at a formal dinner. Saying that some might wonder if he were crazy for "subjecting himself to these constantly yelping village curs instead of just sailing away on his yacht and telling San Diego to go to hell," Spreckels said he could not. "Whatever else I may or may not be, I am not "a quitter.""

Fletcher finally sold the Cuyamaca Water Company, its entire system, and the site for the proposed El Capitan Dam to the La Mesa, Lemon Grove and Spring Valley Irrigation District in 1926 for $1,201,980. In his *Memoirs*, Fletcher stated bitterly that "it cost the City of San Diego at least seven or eight million for its foolish errors in not having accepted earlier proposals to buy the San Diego Flume." He also wrote: "I only made $78,000 in the management and sale of the Cuyamaca System, working over a period of 15 or more years. It was the greatest worry and trouble of my life. It came within an ace of breaking me financially."

The City of San Diego finally moved ahead in 1924 on a plan to build the El Capitan Dam on the San Diego River at the site sold by Fletcher to the La Mesa District. Fletcher and Spreckels fought publicly over the planning and financing for the dam. Fletcher opposed and Spreckels favored the bond issue needed to raise money for the dam because Fletcher wanted the City to accept his proposed compromise that would have given the La Mesa District priority over the river water but assuring plenty for the city of San Diego. The compromise failed and voters approved the $4.5 million bond issue for the El Capitan Dam by a three-to-one majority.

SAN DIEGO'S COMFORTABLE SUPPLY OF WATER

The El Capitan project, designed to solve all of San Diego's water supply problems, was unfortunately stymied for years by a lawsuit brought by Fletcher and the La Mesa, Lemon Grove and Spring Valley Irrigation District. They claimed prior rights to the San Diego River water based on Fletcher's original riparian rights under the old "law of the river" — the "first in time, first in right" doctrine. The City of San Diego responded with its traditional claim to all the water of the river, citing the old "pueblo rights" doctrine that had been regularly upheld by the California courts as giving to a community water-ownership rights on adjacent rivers. Finally, the long-standing conflict between pueblo rights and riparian rights came to a head before the California Supreme Court.

In a dramatic 1930 decision that many thought turned California water law upside down, the Court handed the City of San Diego a resounding victory: "San Diego...is the owner in fee simple of the prior and paramount right to the use of all the water (surface and underground) of the San Diego River." The court reasoned that the pueblo rights doctrine had been transported into American law by the Treaty of Guadalupe Hidalgo which included a general clause confirming pre-existing principles of Mexican law relating to land and water rights. The riparian-rights advocates of "first in time, first in right" were bitterly disappointed by the decision.

The La Mesa, Lemon Grove and Spring Valley Irrigation District did derive some comfort from the recognition by the Court that the district could continue to utilize water "not needed" by San Diego. San Diego definitely did not like this dilution of its absolute ownership right over river water. Meanwhile, the City finalized its plans to build El Capitan, but was stymied one more time by the fact that the La Mesa District owned the land, and had filed a legal action to enjoin construction of the El Capitan Dam. A compromise was finally negotiated that basically gave each side what it wanted. The La Mesa District capitulated on the main issue, expressly recognizing the City's paramount rights over the waters of the San Diego River. In return, the City guaranteed to the La Mesa District all the water needed for its operations. Satisfied with this result, the district turned over all its rights to the land at the dam site.

Construction finally began on the El Capitan Dam in 1932 and its backers enthusiastically prophesied that it would solve San Diego's water supply problems for all time. Like some of its predecessors, El Capitan was designed as a hydraulic fill dam with rock-fill embankments on front and back. Diversion of river water during dam construction was made through a concrete-lined 25-foot-diameter tunnel that was 1200 feet long. Before construction of the final dam, "toe walls" (smaller concrete arch dams about 50 feet high) were built upstream and downstream of the dam location to prevent the slides that can result from hydraulic-filled dams. Walls for the dam itself were built with rock blankets or layers that enclosed the toe walls and decreased in thickness to the crest of the dam. The central portion of the dam was created by hydraulic fill. Earth-fill material suspended in water was sluiced under pressure through large water nozzles into the dam core where the water was allowed to settle out, leaving the fill in place.

Rising 217 feet above the streambed, the dam's foundation was 25 feet below that. The dam was 1240 feet thick at the base, narrowing to 26 feet at the crest. Just over 1000 feet long, it contained nearly 2.7 million cubic yards of earth and rock. Built at a cost of $5.8 million, the reservoir behind the dam could yield 10 million gallons of water per day.

WATER'S ALWAYS "OVER THE NEXT HILL"

When El Capitan was completed, San Diego County was comfortable with its water supply, which depended basically on water from three rivers: the Cottonwood-Otay River (Spreckels' Mountain Water Company system), the San Diego River (Fletcher's Cuyamaca Water Company system), and the San Dieguito River (Fletcher-Henshaw system). With reservoirs on all the major San Diego County rivers, people asked, why the continual outcry about San Diego not having enough water? Actually, most of the reservoirs were kept less than half full much of the time. The usual practice was to build up a reserve water supply in the short rainy season and then allow the reservoirs to be depleted for the rest of the year.

But, said the planners as the years passed, the reserve supply was still not enough. "All those reservoirs" provided by "all those dams" on "all those rivers" would not suffice; the "window," they said, was seven years. There had to be sufficient water storage capacity to last at least seven years. As the "Roaring Twenties" developed, those who ran the figures reported that San Diego's margin of survival was far less than seven years. Actually, it barely reached four years because of the demands of a rapidly growing population which continued to exceed expectations.

So, a critical few among the City's water engineers began looking for another water supply. But the only large-quantity source was the Colorado River, 200 miles east of San Diego, with a range of 6000-foot mountains and 60 miles of desert in between. To all but dreamers, that river seemed inaccessible. Further, the Colorado River water had been the focus of the entire seven-state

El Capitan Dam   *San Diego County Water Authority*

Southwest region (plus Mexico) for a number of years so San Diego would definitely be a latecomer to that big-time water game.

The City ultimately just let the planners dream. Years later, San Diego paid the price for being not just a latecomer but also the last comer to the Colorado River; it became "low man on the totem pole" for rights to that river's water. The City just got along with what it had, believing as Harry Griffen, a prominent water official, recalls in the unpublished *Harry Griffen Manuscript* (on file at Helix Water District archives) that the County's water supplies were "in fairly good shape."

Robert H. Boyle wrote in *The Water Hustlers* (San Francisco: Sierra Club, 1971) about the frontier ethic regarding water, "When it is used up there is more of the same over the next hill." Parched desert travelers were often told by their guides, anxious to keep their thirsty and weary wards moving, that "water will surely be just over the next hill." Those words became a desert mantra repeated so constantly that it acquired both a cynical meaning to pessimists and an expression of hope for optimists. Phrased differently, the mantra was repeated by the politicians and empire builders as, "Don't worry. We'll find a way to get the water." In other words, "water's always over the next hill." That was the comfortable, lulling rationale that prevailed in San Diego right up to World War II.

CHAPTER 3
THE RAPE OF OWENS VALLEY

The population explosion of the late 1800s hit Los Angeles much harder than San Diego. Los Angeles was a railroad terminus for three major transcontinental lines that, with cut-rate fare wars, brought many thousands of people to the Los Angeles area during the boom years. So many that, coupled with the agricultural water demands, the inability of local rivers to meet the public's demand for water was recognized much sooner than it was in San Diego.

Among the first to perceive the looming water problem was William Mulholland, a paradoxical character with a "booster mentality" who started as a penniless Irish immigrant digging ditches for the Los Angeles Water Company and went on to become Superintendent of the Los Angeles water system, the personal embodiment of Los Angeles water policy, and a legend in his time. In the summer of 1904, Superintendent Mulholland, called "Chief" by friends and workers alike, sternly warned his City that "it is probable that there will never in the future be enough water," thus becoming the Los Angeles "Paul Revere of water" who readied Los Angeles to go "over the hill" to find new sources of water supply.

ACQUIRING THE OWENS RIVER FOR LOS ANGELES

Mulholland's chief engineer and close friend, Fred Eaton, became the number one seeker for a plentiful water source that could be tapped for Los Angeles, finally finding one up in the foothills of the Sierra Nevada. Few believed his improbable-sounding promise that this water could actually be brought to the city over the Mojave Desert. Learning about Eaton's find, the no-nonsense Mulholland caustically commanded, "Show me, Eaton." Mulholland later wrote that the two men picked up a "demijohn of whiskey," hired a buckboard with a team of mules, and headed for peaceful Owens Valley at the foot of the Sierras, 250 miles northeast of Los Angeles. As one writer, Mark Wheeler of *Smithsonian*, put it, "despite the hooch, it was the water not the whiskey that made a believer out of Mulholland." "There, Chief," said Eaton, pointing to the Owens River, "is enough water for a city of 2 million people."

Very quietly, to avoid driving prices up, a few men were sent by Mulholland and Eaton to Owens Valley to acquire options on land and water rights that could be obtained quickly and cheaply by what appeared to be small-time speculators. When they were done, Los Angeles had options on all necessary land plus rights to 95% of the Owens River water that flowed through the valley. Los Angeles had secretly acquired, literally, the right to buy a river. Dramatically, the *Los Angeles Times* broke the story with a glaring headline, "TITANIC PROJECT TO GIVE THE CITY A RIVER," that shocked the people of Owens Valley.

Mulholland announced that he would personally supervise construction of the 223-mile Los Angeles Aqueduct needed to bring the water across the Mojave Desert to the San Fernando Valley. This soon became known as "Mulholland's Ditch" and was built over five years at a cost of $23 million. Meanwhile land values fell sharply in Owens Valley and residents were further enraged when it was revealed that a small syndicate of land investors (including owners of the *Los Angeles Times* that had supported the project), could profit handsomely when the aqueduct delivered water to their vast, arid landholdings in San Fernando Valley.

Mulholland, widely hailed for successfully completing the massive project, was urged to run for mayor. "Gentlemen," he responded at a grand ballroom gathering, "I would rather give birth to a porcupine backwards than be mayor of Los Angeles." Then, at the public ceremony attending completion of the project, after many glowing introductions, Mulholland growled simply, "There it is. Take it."

Matters settled down for a while but trouble erupted when disgruntled Owens Valley residents resorted to violence. A caravan of cars with about forty men drove to the aqueduct and dynamited the concrete canal while others opened the spillway gates to send the water back into the Owens River. Mulholland responded quickly, repairing the damage and sending trainloads of city detectives armed with Tommy guns and Winchester carbines to guard the aqueduct. Finally, unable to stand the adverse publicity any longer, Los Angeles virtually "bought out" the opposition by paying in total every claim presented, a grand total of $16 million. A cheap price, some said, for the water bonanza that the City had acquired.

COLLAPSE OF SAINT FRANCIS DAM

But the worst was yet to come for Mulholland. Ominous cracks developed in the Saint Francis Dam built to hold water for the aqueduct. Mulholland personally came out to inspect the dam and publicly pronounced it "sound." That very night at midnight, the dam collapsed and a hundred-foot wall of water swept down the valley, taking over 500 lives, destroying 1250 houses, and inundating 7900 acres of rich farmland. It was called the worst California disaster since the San Francisco earthquake of 1906. Mulholland publicly hinted at possible sabotage, but there was no evidence to support any such charge. Disgraced and visibly broken in spirit, Mulholland resigned and lived out his remaining years quietly as a virtual recluse, a bent and stooped old man.

The Owens Valley story was nationally reported at the time and portrayed in various media with colorful detail. Examples are W.A. Chalfant's *The Story of Inyo* (1933) and Remi Nadeau's engrossing and factual book, *The Water Seekers* (4th ed. 1997). Nadeau called Chalfant's account a "terrific diatribe against Los Angeles." Perhaps Marc Reisner, author of *Cadillac Desert,* had the Owens Valley episode in mind when he famously wrote that in California, "water flows uphill toward power and money." Roman Polanski's 1974 movie, *Chinatown,* starring Jack Nicholson, picks up factual elements of the Owens Valley story including the involvement of Mulholland and Eaton but weaves them into a fictional dark tale of murder and incest. Friends of Mulholland applauded when his daughter Catherine wrote the book, *William Mulholland and the Rise of Los Angeles*, which painstakingly and factually exonerated her father from much of the blame for the Owens Valley disaster.

When it all ended, Owens Valley was for many a dismal landscape, never again to be the tranquil, abundant valley that provided a quiet and plentiful life for its residents. The water had gone south and what some have called a "dust bowl" remained. A recent article in the magazine *Smithsonian* retelling the story describes the valley as "a vast, dusty, cracked-white patch of high desert." Ever since the Los Angeles water project ruined numerous people in Owens Valley, folks there have pronounced over and over the judgement rendered by many residents at the time, "The rape of Owens Valley."

The Owens Valley experience has poisoned the public's minds against thirsty cities including San Diego seeking to obtain more water from agricultural areas, attempts that will be described in ensuing chapters.

CHAPTER 4
MOVING THE COLORADO RIVER WEST

Perhaps the best short description of the storied Colorado River is in the 1916 *Report of the U.S. Geological Survey*:

When the snow melts in the Rocky and Wind River Mountains, a million cascade brooks unite to form a thousand torrent creeks; a thousand torrent creeks unite to form a half a hundred rivers beset with cataracts; half a hundred roaring rivers unite to form the Colorado, which flows, a mad, turbid stream, into the Gulf of California.

Impenetrable canyon walls and protective Ute Indians kept the course of the Colorado River secretive until a one-armed Civil War veteran, John Wesley Powell, came west in 1869 to fulfill an ambition acquired as a young geology professor. A Union major who had lost his right arm at the battle of Shiloh, Powell was fascinated by the West. He eventually became perhaps the single most important figure in Colorado River history. Powell's dramatic reports of his adventurous trips unlocked secrets of the Colorado River, and widely read follow-up stories heightened national interest in the Rocky Mountain states.

Nearly 25 million people are today dependent in one way or another on the Colorado River. With a watershed area of 246,000 square miles, it supplies usable water for seven states. The Colorado River compact of November, 1922, divided the waters between the Upper Basin and the Lower Basin. The Upper Basin states are Colorado, New Mexico, Utah, and Wyoming. The Lower Basin states are Arizona, California, and Nevada. All jousted with Mexico, which is vitally interested in the lower Colorado and the delta area where it joins the Gulf of California.

The Colorado River has been called the "American Nile" because of its impact on agriculture; Marc Reisner in his *Cadillac Desert* calls it "the most controlled, litigated, domesticated, regulated, and over-allocated river in the history of the world." As a "navigable river," it is subject to federal control as a matter of constitutional law; a jokester once said that the definition of navigable water is "any stream that can float a Supreme Court decision." Clearly, the Colorado is a navigable stream for a good portion of its length and the federal government has vested regulatory rights over the river in the U.S. Department of the Interior.

Colorado River   *U.S. Bureau of Reclamation*

THE FIRST TAPPING OF COLORADO RIVER WATER

It was the need for irrigation water that initially opened up the Colorado River. Land in southern California to the west of the river (known as the Colorado Desert) was taken up by early settlers who perceived the farming value of the rich soil formed over centuries by the river's meanderings. The land was much more valuable than it appeared but it desperately needed water through irrigation. A man named George Chaffey saw the potential when settlers from the Midwest and East continued to pour into southern California. His company raised the necessary money and, a masterful promoter, he quickly changed the name of the area west of the river (then a part of San Diego County) from Colorado Desert to Imperial Valley.

Chaffey's operation (California Development Company) tapped the Colorado River in 1901 by building a wooden headgate known as the "Chaffey Gate" 500 feet north of the Mexican border to divert water into the old Imperial Canal, known in earlier years as the Alamo-Imperial Canal because for much of its length it followed the Alamo River riverbed. The canal ran south into Mexico and then west to bypass the towering sand hills west of Yuma, Arizona. Finally, the canal turned north across the border into the "land of the American Nile." Soon, 400 miles of main canals and lateral branches were built across 100,000 acres of the fertile soil. Settlers arrived almost daily as new towns named Imperial, Brawley, Calexico, and El Centro sprung up and the Southern Pacific Railroad served the area. Then, disaster struck in 1905.

THE ALAMO-IMPERIAL CANAL RUNS AMOK

The Alamo-Imperial Canal was vital not only to farmers in Imperial Valley but also to those in northern Baja California, where a syndicate dominated by General Harrison Gray Otis and Harry Chandler, owner and publisher of the powerful *Los Angeles Times*, had bought 380,000 acres of land just across the border. What was called "Chandler's Land" was irrigated with Colorado River water from the canal.

Unfortunately, in 1904 the canal began to accumulate substantial silt at its eastern starting point near the Colorado River. Water flow diminished seriously and farmers demanded action to open up the canal. To gain time, workers cut a new 60-foot intake in the bank of the Colorado just 4 miles south of the border. The cut allowed river water to flow directly into the canal, bypassing the silted-up portions of the canal to the north.

When floodwaters broke a temporary control gate at the cut in 1905, water — far too much water — gushed into the canal, widening the cut as it came through the bank. Efforts to close the cut failed, the capricious Colorado took over, and the cut quickly widened from 60 feet to 600 feet. No canal could contain the water, as practically the whole of the Colorado River poured through to cover the land all the way to the Salton Sink, a vast depressed area below sea level that had once been ancient Lake Cahuilla.

When the flood poured over an embankment into the Salton Sink, it created a waterfall as much as 28 feet, and this by its action created the phenomenon known as a "cutback." The land-carving action of the waterfall against the lip of the embankment eroded the land and forced the bank to move steadily backward. Upstream it went across Imperial Valley, reaching the amazing pace of 4000 feet a day, creating a roar that could be heard for miles. Wiping out farms and homes in its path, the traveling cutback threatened the towns of Calexico and Mexicali. Finally, imaginative engineers with hundreds of courageous laborers working around the clock with strong doses of dynamite turned the cutback into a circle much the same way that cowboys forced stampeding cattle into a circle to slow down the herd. This broke the flowing water into a series of little cutbacks that slowly died away.

Meanwhile, massive manpower was thrown into the job of closing the cut on the Colorado River. Rails were laid by the Southern Pacific on hastily thrown-up trestles built on 90-foot pilings pounded by huge steam-driven pile drivers through 30 feet of rushing water deep into the riverbed to reach a solid footing. At one point, a telegram went to Southern Pacific headquarters: "We have exhausted all available supply of piles in San Diego and southern California."

The Southern Pacific's entire freight car system was placed at the disposal of the construction engineers.

Over 3000 flatcars and "battleship" dump cars were loaded at quarries along the railroad's route and rushed to the flood scene to dump thousands of tons of rock, gravel, and clay into the breach. The frantic action went on for months until the cut was finally closed in February 1907, and the river was forced back into its banks. The 50-mile-long, 15-mile-wide body of water formed by the 1905 flood in Imperial Valley was named the Salton Sea, and in 1909 the state carved out of the old San Diego County a new Imperial County.

The old canals resumed operation and, with its water supply protected, Imperial Valley began to produce an agricultural abundance. Some locals said (exaggerating, as boosters will) that Imperial Valley was becoming a "garden for the world."

THE ALL-AMERICAN CANAL IS BORN IN IMPERIAL VALLEY

The Imperial Irrigation District (known as IID) was formed in 1911, and by 1916 had acquired all the assets of the old California Development Company, including all the canals and levees in the valley, so that finally the valley people (through the district run by elected officials) controlled their water supply. But that supply was not reliable because of occasional seasonal flooding from the Colorado River and possible interference from Mexico (also served by the old Alamo-Imperial Canal). Some valley farmers were content with the status quo, as was the Chandler syndicate on the Mexican side of the river, but many in Imperial Valley wanted a new canal built solely on the American side of the river (the All-American Canal, as it was called), free from Mexican interference. The *San Diego Union* strongly favored the project but leading the opposition was the powerful voice of Chandler's *Los Angeles Times*.

In 1919, the Imperial Irrigation District sent its young attorney, Phil Swing (later to become a member of Congress) to Washington to plead its case for the new canal and storage reservoir on the Colorado River to ensure a stable flow of water into the canal. That meant a new dam. Thus began the huge Colorado River project that would require massive dams generating electricity to pay for the project and an aqueduct to carry river water to the Pacific Coast. In opposition were the Upper Basin states, private power companies not wanting the public power competition, and the state of Arizona.

San Diego took steps to get its nose under the Colorado River tent. Shelley J. Higgins, city attorney, drove with his deputy over the unpaved desert highway to the site on the river where water would be diverted for the projected All-American Canal. Following the gold miners' tradition, Higgins erected a small rock monument and placed inside a can containing a legal claim to river water on behalf of the people of San Diego. William Jennings, a San Diego water attorney, recalled later in his *William H. Jennings; Water Lawyer* (available at the San Diego Historical Society archives) that many in San Diego called these men dreamers, laughing that any thought of San Diego getting water out of the Colorado River was "as remote as finding a pot of gold at the end of the rainbow." The Higgins action became known as the "claim in a can" and was formalized when, at the urging of Fred Heilbron, a written claim for 112,000 acre-feet per year of Colorado River water was filed with the Department of the Interior in Washington.

In 1922, a new Colorado River Commission was formed, made up of representatives of the seven states in the Colorado River basin with Secretary of Commerce Herbert Hoover acting as chairman. Under his capable leadership, the Commission hammered out the historic Colorado River Compact which apportioned Colorado River water between the Upper and Lower Basin states.

Each basin was apportioned "in perpetuity" 7.5 million acre-feet per annum. The Lower Basin was given the right to increase its use by one million acre-feet. The Upper Basin agreed not to deplete the flow below an aggregate of 75 million acre-feet over ten consecutive years — or, on average, 7.5 million acre-feet per year.

(Many years later, the Upper Basin, by agreement in 1948, divided its annual rights — Colorado, 51.75%, New Mexico, 11.25%, Utah, 23.00%, and Wyoming, 14.00%. By a 1964 Supreme Court decree, the annual rights of the Lower Basin were divided — California, 4.4 million acre-feet, Arizona, 2.8 million acre-feet, and Nevada, .3 million acre-feet.)

Congress approved the compact and legislatures of six of the seven basin states, including California, ratified it. Arizona, however, threw a monkey wrench into the proceedings and for years led the fight against the enabling legislation (known as the Swing-Johnson Bill) sponsored by the new Congressman Swing from Imperial Valley and California's Senator Hiram Johnson. The *Los Angeles Times* joined Arizona in fighting the legislation; some said that its opposition stemmed from the publisher's massive land holding in northern Baja California where the All-American Canal was strongly opposed, and from the newspaper's conservative opposition to public power projects. One prominent "Zonie" (California slang for a person from Arizona) opposed the project with a widely distributed pamphlet that screamed with little logic, "Construction profit goes to Las Vegas; the franchise goes to Nevada; the power goes to Los Angeles; the water goes to Mexico; and Arizona goes to hell."

In six successive sessions of Congress, year after year, the two dedicated legislators reintroduced their proposed bill and five times they were defeated, often by filibusters led by Arizona's senators. Finally, enough votes were mustered to break the filibuster and after the Senate approved the legislation the

House of Representatives quickly followed suit. On December 21, 1928, President Calvin Coolidge signed the Swing-Johnson Bill into law. In 1931, the Supreme Court threw out the Arizona lawsuit seeking to kill the legislation. Finally, work could begin on the massive project.

Essential elements of harnessing the Colorado River were Boulder Dam (to store water and generate electric power), Parker Dam (to create the reservoir needed to stabilize flow of river water for diversion), Colorado River Aqueduct (to take the diverted water to the West Coast), and the All-American Canal (to supply water to Imperial Valley). There was no precedent. In design challenge and construction magnitude, the projects made engineering history both in the United States and around the world. Much-needed employment during the depth of the Great Depression would be provided for thousands of men and women, and the massive cost in federal dollars could be recovered through the sale of power generated at Boulder Dam, later renamed Hoover Dam.

CONSTRUCTION OF THE ALL-AMERICAN CANAL

Because the All-American Canal had started the progression of projects that culminated in the package embodied in the Swing-Johnson Bill, Representative Phil Swing from Imperial Valley often remarked in later years that the Canal was "the tail that wagged the dog." This was the canal that the San Diego City engineering department hoped would bring west San Diego's share of the Colorado River water.

In Washington, the usual delays ensued and Phil Swing, anxious to get the project started, finally took his case to President Franklin D. Roosevelt soon after the new president took office in 1933. Pleading that his legislation would create thousands of jobs, Swing finally got the cherished appointment — fifteen minutes, he was told, and "don't talk too long." Swing spoke for just ten minutes and Roosevelt, ever the good politician, responded amiably but without commitment and put in a call to Senator Hiram Johnson of California. Johnson had deserted his Republican party the year before to support Roosevelt against Hoover. Johnson, fortunately, gave unstinting approval to the project.

The very next day, Swing was told to come to the Public Works Administration to help with the paperwork to get the money flowing. A jubilant Swing wired Imperial Irrigation District: "Glad advise canal approved and six million allotted start work."

In 1935, about 20 miles northeast of Yuma, work began on the Imperial Dam, a first for the world of irrigation that would allow the Colorado River water to be diverted into a giant desilting plant before it entered the canal. The canal, about half a city block wide and 22 feet deep, ran south (parallel to the river) almost to the border and then due west in almost a straight line for over 50 miles. Surveyors and rodmen, their assistants, painstakingly staked out

All-American Canal   *U.S. Bureau of Reclamation*

Imperial Dam   *U.S. Bureau of Reclamation*

the route; following them were workers who cut away the tough mesquite with sharp, hooked tools that looked like medieval weapons. Tractors then broke the hard crust on the soil and loosened the earth. In 1936, an article on the All-American Canal appeared in *Popular Science* magazine. Entitled "The World's Biggest Ditch," the article described in gripping terms the "army of workmen gashing the desert through miles of sand, sage and mesquite."

Frequently used to carve the ditch and remove tons of dirt were four-mule teams pulling what were called "fresno" scrapers manned by unemployed farm-hands who were paid three dollars a day, decent Depression-era wages. The work was relatively straightforward until the much-feared pure desert was encountered. There, dunes of white sand piled high and often moved by the wind (they were called "walking dunes") were difficult to penetrate.

Huge dragline machines called "walking cranes" were brought in to cut through the dunes; 650 tons of machinery that required 20 boxcars to carry the parts to a railroad siding near the construction work. Wheels would only sink into the soft sand, so the mammoth machine operated by 8 men had 2 mechanical "feet" mounted off-center on an axle. As the axle turned, the 21-ton feet actually "walked" across the sand taking 7 feet per step. A long neck extended out half a block in length, controlling a huge digging bucket big enough to hold an automobile. Each scoop filled the bucket with 50,000 pounds of dirt that was then swung over and dumped on the embankment alongside the ditch.

The machine "walked" along a 20-foot-wide roadway built on the side of the canal. Not all dirt went onto the embankments. Dirt needed to fill low spots was dumped into tremendous buggies that rode on balloon tires 8 feet in diameter. Loose dirt on the embankments was compacted with "sheep's foot" tampers, large rollers studded with iron "feet" and pulled by huge tractors. Finally, oil or vegetation was used to stabilize the embankments. Overall, the 60 million cubic feet of material moved during the project could fill a trainload of standard gravel cars extending for 2235 miles, the distance from Los Angeles to Chicago.

After six years of hard and wearying work plus $25 million the Colorado River water flowed into Imperial Valley where IID distributed it to thirsty farms.

COACHELLA BRANCH COMPLETED
AND THE VALLEY BLOSSOMS

The Coachella Branch, extending northwest from the main canal below Salton Sea, was built a few years later to provide a water supply for the vast agricultural acreage on what was called the East Mesa of the California desert. This branch of the "Great Ditch" ran north on the east side of the Salton Sea and then circled westward above and around the sea in the Coachella Valley. It supplied another huge agricultural area and helped develop Palm Springs, just a few miles north. The Colorado River water brought even more development to Imperial Valley. Once called the "Valley of the Dead," it now brought to productive life more than half a million acres that eventually yielded nearly 1 billion dollars in crops.

On October 1, 1934, a contract was finalized by San Diego with the U.S. Interior Department and the City of San Diego to carry San Diego's 112,000 acre-feet annual allotment of Colorado River water west through the All-American Canal. In 1937, the City obtained the Hill-Ready-Buwalda engineering report on moving the water over the mountains to the coast. Water would be conveyed from the canal to the Borrego Valley. Then, a high-powered pump would lift the water 2700 feet above sea level to a point where, through a 7-mile tunnel, the water could be carried through the Cuyamaca Mountains just south of Julian and then down through pipes to the El Capitan Reservoir. It would be a four-year task costing an estimated $8.65 million. The project was hailed by the *San Diego Union* in February, 1931, as "Bringing Rocky Mountain waters to develop our 'Harbor of the Sun.'" But no action was taken on the project and it has never been built.

BUILDING OF BOULDER (LATER HOOVER) DAM

(Author's Note: A frequent source for human interest material dealing with the building of Boulder Dam, Parker Dam, and the Colorado River Aqueduct is the previously described *Harry Griffen Manuscript*.)

Boulder Dam was designed to conquer the Colorado River and make possible the river diversion that would supply water for southern California. In charge of building it was the Six Companies, Inc. construction consortium, so named for the six firms that became involved through competitive bidding. Dominant personalities in the consortium were Henry J. Kaiser, a flamboyant 48-year-old contractor from Oakland who became chairman of the board and Frank T. Crowe, the project's superintendent of construction. Harry Griffen, in the *Harry Griffen Manuscript*, labeled Crowe "America's foremost dam builder" and described him picturesquely: "With his slouch hat, slide rule and faded khakis, the lanky wind-burned Crowe was the prototype of the hard-driving, two-fisted, field engineer featured in film and fiction." Crowe called the project "the biggest dam ever built by anyone anywhere."

The federal government allotted seven years for construction of the dam. Crowe cut that to five and substantially reduced costs with his plan to light the dam site and have shifts work around the clock seven days a week, "24/7" it is now called. Cheap labor, Crowe said, would not be a problem since announcement of the project during the depth of the Depression would bring thousands of workers. Crowe was correct and people flocked to Nevada, often camping in and around Las Vegas while they waited for jobs. The Bureau of Reclamation approved the final Six Companies contract at $49 million.

Black Canyon became the final site for the dam. Sheer canyon walls dropped 800 feet to the river level. Access to the construction site was difficult, and everything that was needed had to be brought in from the outside. Tunnels had to be driven into unforgiving rock to divert the river water until the dam could rise to the water-containment level and thereafter to act as flood-time spillways. Tractor-driven rigs, each mounted with 24 jackhammers, bored holes for dynamite placement and after the ear-shattering blasts, mucking crews swarmed in to shovel the debris into trucks. This was backbreaking work in hellish conditions. Temperatures sometimes soared unbearably, causing heat prostration, and carbon monoxide poisoning took the lives of thirteen men.

Far above these workers and given the job of clearing loose rock from the canyon walls were men swaying in bosun's chairs from ropes dangling from the Arizona and Nevada canyon rims. Any slip meant a death dive into the river far, far below. Highscalers, these men were called, and they had perhaps the most dangerous jobs on the dam.

Once the Colorado River was diverted through the bypass tunnels, the riverbed had to be cleaned all the way down to bedrock where the dam could be anchored. Steam shovels and cranes fed on the muck, a ton at a gulp. Working for six months on a 24/7 basis, swarms of men and machines finished stripping away layer after layer after layer of the Mesozoic ooze and primeval muck that had taken the river eons to create.

Finally, actual dam construction began. Buckets containing 8 tons of wet concrete were swung down from the Nevada rim, one every minute, and dumped into house-sized wooden forms, month after month for two years. Gradually, up and up rose the concrete dam, looking like a collection of giant building blocks. The engineers knew that those blocks would gradually take the form of a giant horizontal arch; in size tremendously greater but in basic shape the same as the Sweetwater, Barrett, Otay, and Morena dams.

When completed, Boulder Dam would be 660 feet thick at the base, narrowing to 45 feet thick at the top and extending 1244 feet across Black Canyon. A grand 726.4 feet high, it would weigh over 6.5 million tons. Some compare it as an engineering achievement with the Pyramids or the Sphinx, perhaps to be eclipsed only by China's massive Three Gorges Dam or India's Sardar Sarovar Project.

Unfortunately, partisan political controversy bedeviled the dam's name for years. From its inception, those involved had called it Boulder Dam and that name took hold even though it was built in Black Canyon, not Boulder Canyon. But Interior Secretary Ray Lyman Wilbur named it Hoover Dam after his boss, President Hoover, when he drove a silver spike in a railroad spur at Las Vegas when preparations were being made to build the dam. Congressman Swing, a Democrat who had shepherded the enabling legislation through Congress, remonstrated volubly. He recognized Hoover's leadership in forming the

Black Canyon   *U.S. Bureau of Reclamation*

Hoover Dam under construction   *U.S. Bureau of Reclamation*

Hoover Dam   *U.S. Bureau of Reclamation*

Colorado River Compact but he thought Hoover had not been particularly helpful in the long and acrimonious legislative battle that ensued.

When the Democrats took over in 1932, incoming Interior Secretary Harold Ickes killed the Hoover name and restored "Boulder Dam," which it remained for 14 years of Congressional rule by the Democrats. Then, in 1947, the first Republican Congress to come along renamed it "Hoover Dam" and President Harry Truman, although a strong Democrat, made it bipartisan by signing the bill without comment.

Construction of the dam was costly in terms of human life. Heat prostration, falls, explosions, drownings, and other accidents killed 110 workers during the five years of construction. Interestingly, J.G. Tierney and P.W. Tierney, father and son, were the first and last to be killed. There is no count of those injured. But despite these tragedies, *esprit de corps* was always high; "there is something peculiarly satisfying about building a great dam," said Superintendent Crowe. Theodore White of *Harper's* reported with perhaps understandable exaggeration that the men worked "with hilarious looseness." To them, the dam assumed a personality of its own. On March 23, 1935, more than two years ahead of schedule, Superintendent Crowe declared Boulder Dam finished. Hoover Dam has been its name now for over half a century.

PARKER DAM IGNITES THE ARIZONA-CALIFORNIA WAR

The Los Angeles-based Metropolitan Water District (MWD) had been given statutory authority when it was created in the 1920s to build and maintain an aqueduct to the coastal plain, including specifically the counties of Los Angeles, San Bernardino, Riverside, Orange, and San Diego. The consortium of contractors that bid successfully on the aqueduct was again named "Six Companies."

The first step had to be a storage reservoir that would control the flow of water into the aqueduct and make it independent of the rises and falls of the Colorado River. To create that reservoir, Parker Dam was designed to be built 155 miles south of Boulder Dam. To be anchored in a canyon of the Whipple Mountains, it would be the "deepest" dam in the world, requiring diamond drilling through the layers of rock that formed the riverbed and then excavation 240 feet below that to reach the essential bedrock.

Trouble first began when Arizona "drew a line in the dirt" on its side of the Colorado River and said "no dam until our water rights are resolved to Arizona's satisfaction in Washington." That could mean years of delay. Arizona then seized on the fact that Parker Dam was being built by the Bureau of Reclamation without any specific Congressional authorization. "Parker Dam is illegal," cried Arizona's spokesmen, and Arizona vowed that it would stop the project. When the actual construction process neared Arizona on the other side of the Colorado River, Governor Benjamin B. Moeur of Arizona cabled a warning to his counterpart in California: "This could mean war." When heavy cables to hold the river barges in place for riverbed drilling were anchored on the Arizona bank, Moeur declared martial law and ordered militiamen to the site.

The *Los Angeles Times* rushed a "war correspondent" to what it called "the front" and he, tongue firmly in cheek, wrote of the "impending movement of State troops into this theater of war to protect the State of Arizona from invasion by all or part of the State of California." More realistic old-timers did not hold their breath, pointing out that mountain goats would have to be used by any Arizona militiamen trying to reach the scene.

Parker Dam   *U.S. Bureau of Reclamation*

Undaunted, the Governor's aide and the major in charge of the 158th Infantry Regiment of the Arizona National Guard decided to launch an amphibious landing. The Arizona troops, together with the *Times* reporter, were loaded onto the *Julia B.* and *Nellie T.* (a pair of old 45-foot ferryboats powered by Model-T engines that a colorful couple, Nellie and Joe Bush, two leading citizens of Parker on the Arizona side, ran across the river to the town of Earp on the California side). The reporter quickly filed a wire story describing the "Arizona Navy" and the comic opera story moved across the wires to newspapers throughout the nation.

Arizona saw nothing funny about the "California encroachment crisis." Governor Moeur solemnly declared that he was ordering into action whatever troops would be needed to "repel the threatened invasion of the sovereignty and territory of the State of Arizona." Almost unbelievably, Arizona officially petitioned Congress to send the Navy's battleship *U.S.S. Arizona* up the Colorado River to reinforce the *Julia B.* and *Nellie T.* Ignoring all this, the Six Companies crews worked doggedly along, driving pilings across the river closer and closer to the Arizona bank. Reporters, newsreel cameramen and photographers together with miners, cowboys and various other folks came to Parker to see the "war" begin as Joe Bush readied the *Julia B.* to move troops to the site.

But the river level fell and the "Arizona Navy" was thwarted; there was not enough water to float the "fleet."

This did not stop Arizona. Rumbling north out of Phoenix came eighteen army trucks stuffed with over a hundred National Guard troops together with machine guns and a hospital unit. Governor Moeur issued a grim statement: "We may get licked in the affair, but we will go down fighting." With Six Companies adamant about continuing work and the Arizona troops advancing, Interior Secretary Ickes finally had enough and stepped in. "Stop work," he telegraphed. Six Companies complied, work ceased, Governor Moeur declared victory, and the troops were recalled.

The battle was then transferred to the courts. There, Arizona proclaimed, "victory is ours" after the judge agreed that the dam had not been properly authorized by Congress. But Arizona's victory was short-lived. Congress quickly filled the legal gap by officially authorizing the dam, and the ignoble, comic-opera "Arizona-California War" was over.

THE AQUEDUCT INCHES ITS WAY TO THE WEST COAST

There remained the task of moving the Colorado River water over desert and mountains to the coast. For five years, over 35,000 people and innumerable machines labored at building the Colorado River Aqueduct. Most of the workers were paid $5 per day. Still, that was good money at that time and more than 100,000 Depression-weary people applied for jobs at MWD's employment offices. Thousands more sought jobs at the contractors' offices on the job sites.

Among the workers were some real characters. There was Pistol Pete, a former Texas Ranger who carried a .38 automatic and wore a badge as a deputy sheriff for all of Riverside, Los Angeles and San Bernardino counties. He was a slight but powerful man who was an expert at breaking up fights among hard-drinking, brawling men. Bughouse McCabe, a superintendent on several big jobs, actually liked his nickname and carried mental institution release papers as proof of his identity. Doughbelly LaPlante, a bear-wrestler, left a traveling carnival in Indio to join a tunneling crew. Cactus Kate had been the crusty madam of a popular bordello. Doctors, lawyers, and at least one well-known society playboy, all hard up for money, joined the workforce.

The gargantuan undertaking began at Lake Havasu (formed by Parker Dam) and required powerful pumps to take the water over the Whipple Mountains. Then, at over 1000 feet above sea level, the water could be sent through a short tunnel before dropping down to cross the desert through open canals. After that came tunnels through a series of mountain barriers, some having peaks of over 10,000 feet.

The tunneling was frustratingly slow at the beginning; less than 2 miles in over 18 months. Huge amounts of water were required, both for domestic needs by the thousands of workers and for a multitude of construction requirements such as mixing and curing concrete, air compression machinery cooling, and recharging storage batteries. The construction water had to be pumped through tunnels from the Colorado River and averaged about 23 million gallons monthly; the greatest monthly amount was 46.636 million gallons in July 1936.

A new type of equipment was used for the job, creating 29 tunnels that extended for 92 miles. Huge machines holding eleven power drills assaulted the face of the tunnel while "powder monkeys" filled their holes with dynamite. After each ear-splitting blast, large blowers sucked out the gas and the "muckers" cleared loose rock. Slowly, steadily, the grimy crews blasted their way through the mountains, progressing at the rate of about 20 feet a day. The miners were men who "liked their likker" and favored a bar in Indio called Jackhammer. More than one of those hard-driving, hard-drinking men said the Colorado River Aqueduct could not have been built without that Jackhammer.

After the tough tunneling, digging the ditch through many miles of desert was equally challenging and, again, new equipment was used. First to make the "rough dig" came the mammoth dragline cranes (described earlier in the section on the All-American Canal). Then, the invention of one of the Six Companies called the "canal trimmer" swung into action. This behemoth operated on tracks on each side of the ditch, crawling along at the rate of about 1 foot each minute, cutting the rough ditch into the desired canal shape. On its heels was another piece of monster machinery, a "canal paver," which spread a concrete lining over the sides and bottom of the ditch and then tamped it into place. Their large size dictated that these would be "one-use-only" machines.

This was the same desert that General "Blood and Guts" George Patton of World War II fame used to train his army in 1942. The army tapped the aqueduct for water at 15 locations along its desert length. Patton drove them hard, always carrying his trademark pearl-handled revolver on his right hip as he tough-trained the men to confront German Field Marshal Erwin Rommel and his feared Afrika Corps. As many as 200,000 soldiers went through Patton's tough Desert Training Center and survived to call it "the place that God forgot."

Numerous schools were built, depending on the number of workers with families at different times during the years of the construction. Emergency first-aid facilities were established along the aqueduct route, and a 26-bed hospital was built at Indio. Dr. Sidney Garfield had another hospital at a tiny community named Desert Center, a wooden, 12-bed facility. MWD set up a pre-pay system with five cents deducted from workers' paychecks for unlimited medical care from Dr. Garfield. He was called the "nickel a day doctor" and, in addition to providing medical treatment for thousands of workers, constantly stressed preventive medicine and education.

The stories told about people and events involved in building the aqueduct were endless; some funny, some cruel, some tragic. The *Harry Griffen Manuscript* supplied many of the stories for this account. In one, Griffen recounts an occasion when visiting city bureaucrats became concerned about a rumored organization named "The Ancient Order of Aqueduct Billy Goats," which supposedly had secret rites and controversial initiating procedures. Actually, the membership was rather informal but limited to "extremely sunburned surveyors and hairy-eared engineers who had roamed and crawled through cactus, desert brush, sidewinders, and rattlesnakes" when the alignment for the main line aqueduct was established. It was clearly an exclusive group of individuals. No one questioned that only two categories of animals, billy goats and jackasses, could survive this environment. Not wishing to adopt the latter name, the club's surveyors and engineers chose to name their informal society after the billy goats.

Workers with their families were often forced to live in temporary camps out on the "front lines" where men fought constantly with sand, dirt and rock. Stories told would inevitably include the names of the camps: Old Woman's Mountains, Chuckawalla Valley, Edom Hills, Devil's Elbow, Berdoo, Pushawalla, Yellow, Cactus City, Utopia, Jackhammer, Owl, Paradise Cafes, and some so ribald as to be better left unreported. As the tunneling progressed, there were numerous "holing through parties"; these were, after all, "hard rockers," hard-working and hard-drinking construction crews fighting their way through solid rock.

CONQUERING OLD SAN JACK FINISHES THE AQUEDUCT

The Colorado River Aqueduct was featured in the movie *Empire of the West*, starring Leo Carillo. That led to a nationally broadcast *Fox Movietone News* program narrated by Lowell Thomas when the nation first saw the magnitude of the project. The last tunnel of the Colorado River Aqueduct was bored for 13 miles under Mount San Jacinto. This peak in Riverside County is the second highest in southern California. "Indomitable Old San Jack" presented seemingly insurmountable problems. Water faults were encountered that repeatedly flooded the tunnel. Working against the odds and fighting off the floods, a lateral shaft was built that, connected with the main shaft, would avoid the water onslaughts and enable the tunnel to be completed. The two shafts came together late in 1938, five years after the tunneling had started in May 1933. Amazingly, the two shafts met with only a tenth of an inch difference in elevation and none laterally; a remarkable engineering feat. At tunnel end, a CBS radio microphone recorded the last explosion that opened to the setting sun the 13-mile hole through Old San Jack.

After the tunnel had been lined with cement, it was time for the completion ceremony. One of the giant water lines of the world with tunnel and conduit sections 16 feet in diameter and large enough to accommodate a railway loco-motive was ready to deliver water. At 1:20 p.m. on January 7, 1939, about 1000 people gathered at the western portal of the Colorado River Aqueduct and, along with a nationwide radio audience, heard the general manager give the long-awaited order: "You may start the pumps as soon as you are ready." They did. And out of the tunnel roared a raging white torrent of Colorado River water at the rate of 180,000 gallons per minute.

More than 10,000 men and women had worked on the project. At the end, it would take a drop of water 57 hours to travel the 242 miles from the Colo-rado River. To finance that trip, a total of $220 million had been spent, a little under $1 million per mile. The aqueduct finally emptied into Lake Matthews in Riverside County, capable of delivering over a billion gallons of water a day.

CHAPTER 5
FATEFUL DECISIONS FOR SAN DIEGO

After World War II, San Diego came face-to-face with the hard fact that it could not generate enough water to supply its needs. As water lawyer Bill Jennings put it, "We came out of the war with all our water gone." San Diego simply had to reach out for imported water since local supplies would never again suffice. As the need for more and more water grew, San Diego had several alternatives to control for itself those new sources of supply, first from the Colorado River and second from the Feather River.

TWO CHOICES FOR OBTAINING COLORADO RIVER WATER

For years, water engineers seeing the imminent water crisis had the foresight to seriously consider the two choices for collecting San Diego's share of Colorado River water. Build an aqueduct north along the coast to join San Diego with the Colorado River Aqueduct that had been built by MWD (headquartered in Los Angeles and responsible by state law for distributing water imported from the Colorado River). Or build an aqueduct east over the mountains and take San Diego's share of the Colorado River water from the All-American Canal. The City of San Diego had kept alive its contract with the Department of the Interior and the Imperial Irrigation District of Imperial Valley, paying a small annual amount to maintain its right to carry the Colorado River water through the All-American Canal. Bill Jennings recorded the debate over the alternatives of joining MWD to the north or the All-American canal to the east: "We reviewed and studied and debated very strongly the alternate to bring [the water] over the mountains from Imperial Valley."

Many thought this canal choice would be best, fearing that joining MWD would lock San Diego into an eternal second-place position behind Los Angeles. Countering this were the demonstrable facts that the need was urgent and the MWD aqueduct was already there, ready to be tapped with a pipeline running directly south to San Diego. Also, some reasoned, the second-place argument was based too much on emotion. In his *Harry Griffen Manuscript,* Griffen found it "clear at the time that the people of San Diego were hurt about the idea of having to join something Los Angeles had developed." Money considerations argued in favor of the MWD choice (round figures, $17 million versus

$25 million for the canal route). So were time considerations: two years for the former and three for the latter.

Perhaps most important, this was wartime and the federal government was involved because of the vital military installations in San Diego. A presidential study committee was appointed with a request for quick recommendations and one member, Phil Swing, was from San Diego and Imperial Valley, certainly not prejudiced against the All-American Canal alternative. The committee's report favored the northern alternative, advocating federal funds estimated at $17.5 million for immediate construction of an aqueduct from the MWD Colorado River Aqueduct to the San Vicente Reservoir (San Vicente Lake) in San Diego County. This report went all the way up to President Roosevelt's desk.

Time was of the essence to the military. The U.S. Bureau of Reclamation had warned publicly that it would be foolhardy to rely on a continuation of the wet seasons that had predominated in recent years. Already, drier years were commencing and a drought period was being predicted. The Navy clearly favored the MWD solution, swayed by the fact that this would be the quickest way to get an assured supply of water to the critical San Diego military installations and, besides, would require a minimum of wartime-scarce critical materials. President Roosevelt agreed very quickly and in November 1944 sent a message to the Senate recommending approval of the San Diego Aqueduct designed to tap into MWD's Colorado River Aqueduct.

Bill Jennings called this a "take it or leave it proposition, so we took it." The mayor and other officials went to Washington and the deal was finalized. But then a fatal glitch developed. Only project authorization and not appropriation of the necessary money had gone through Congress. The federal lawyers decided that war powers could no longer justify such an expenditure without congressional approval. Jennings recalled that federal authorities "notified San Diego they were going to cancel all bids for the San Diego Aqueduct and drop the planned construction. That meant San Diego would have to start from scratch and float a bond issue. So we put all the pressure on Uncle Sam that we could."

Some commentators have written that President Roosevelt directed that the project go forward, giving San Diego no choice but to go along. But, actually, there was a choice after the war powers issue was raised. Legally, San Diego could have at this point walked away from the project that tied its imported water future to MWD. It did not.

The record is not at all clear as to who made the fateful decision to build the San Diego Aqueduct, if indeed any formal decision was made. A good case can be made that wartime considerations and momentum of events carried the day without anyone pushing hard for San Diego to declare its water independence and proceed with the existing plans for a hookup to the All-American Canal. A new contract may have been necessary since the Secretary of the Interior had veto power over the existing contract and, besides, the contract had a ten-year term that expired in 1946. Also, voter approval would be required for a bond issue to provide the necessary construction costs for the All-American Canal alternative; it is anyone's guess whether such a bond issue would have carried. It is also anyone's guess whether the Navy would have tolerated the delay this would have caused.

The die was certainly cast by events. Attorney Bill Jennings still called it a "forced decision; the federal government decided that they would, willy-nilly, build a line to supply the Naval installations in San Diego County from the MWD aqueduct as the shortest and quickest way to get the water." Mayor Knox went to Washington with a special delegation from SDCWA headed by Chairman Fred Heilbron and Bill Jennings to get the aqueduct project back on the track. For a while, it was touch-and-go. The water situation remained chancy and the bureaucrats were not happy that the Navy wanted to go right ahead and build the aqueduct without a Congressional appropriation.

A convincing figure before the Congressional committees, tall and commanding (6-foot, 5-inch) Fred Heilbron presented San Diego's case forcefully and persuasively. He played skillfully on the theme that the Navy was largely responsible for using up during the wartime years the water that San Diego had carefully stored away in its reservoirs and therefore, as Heilbron often put it "owed San Diego big-time." Heilbron's speaking talents were later recognized by no less than Senator Joe McCarthy who, during Senate committee consideration of a later water pipeline to San Diego, remarked that he was always afraid while practicing law as a private attorney that he would come up against a powerful and persuasive witness like Heilbron.

The final go-ahead deal negotiated by Heilbron was a lease-purchase arrangement. The Navy would pay for the aqueduct and lease it to San Diego; lease payments would be $500,000 annually over a term of 32 years. San Diego would be given an option to purchase at any time at the final price less lease payments already made. Regrettably, San Diego got a 6-foot-diameter pipeline instead of the 7-foot-diameter one that would have carried all of the Colorado River water to which San Diego was entitled under the existing apportionment. This was because the Navy's plan for the aqueduct designed it to carry just enough water for the military installations in San Diego.

But there were more hard pills to be swallowed. MWD took the tough position that San Diego could not transport its Colorado River entitlement through its Colorado River Aqueduct until it agreed to SDCWA being annexed to MWD. San Diego objected strongly but finally agreed; indeed, it had little choice. Bill Jennings favored the move, arguing that the objections were "sort of parochial and provincial."

As finally negotiated, the agreement committed San Diego to pay an annexation charge of $13 million spread over a 30-year period without interest going forward. The total was calculated as the amount SDCWA would have paid to MWD in tax money if it had been a member agency since MWD was organized in 1928, plus interest. In effect, some said, it was San Diego's share of the cost of constructing the Colorado River Aqueduct. One of the annexation terms was that MWD would pay half the cost of building the pipeline that would bring the Colorado River water to San Diego County, with MWD obligated to operate and later own the northern half of the pipeline. That is why the "point of delivery" for the water to SDCWA is 6 miles south of the San Diego-Riverside county line, causing problems between SDCWA and MWD to this day.

Reasonable people can argue whether the annexation was a better deal for MWD or San Diego. MWD and the City of Los Angeles certainly benefited hugely. MWD had just incurred very substantial indebtedness to build the costly Colorado River Aqueduct and at the beginning it had virtually no customers for the water. Cash flow was essential and San Diego had the cash; in fact, it bought over half the water that MWD had to sell in its early years.

The City of Los Angeles still likes to point out that it supplied most of the money to build the aqueduct because of its predominant position on real estate tax assessments, and to this day argues that it heavily subsidized the facility that made Colorado River water possible for San Diego. Thus were seeds planted long ago for the discord that has subsequently grown between San Diego and Los Angeles over water supply issues.

The City of San Diego was not totally happy in 1945 since another part of the deal was the formation of the San Diego County Water Authority (SDCWA), and that meant giving up some of the City's control over imported water. In the end, the City of San Diego became a part of SDCWA and SDCWA became a member agency of MWD. San Diego's Colorado River rights were merged into those of MWD. The SDCWA-MWD annexation proposition was submitted to and approved by the voters; the ayes prevailed by a ratio of 14 to 1.

The new San Diego Aqueduct could now be built. It would be a gravity conduit tapping into MWD's Colorado River Aqueduct at the western end of the Old San Jack (Mount San Jacinto) tunnel in Riverside County and running south to the San Vicente Reservoir, the storage-supply reservoir for San Diego County. Total length would be approximately 70 miles, with 30 in Riverside County and 40 in San Diego County. The pipeline would be 6-foot-diameter concrete. Speed in construction was viewed as essential in view of San Diego's dwindling local water supply.

BUILDING THE SAN DIEGO AQUEDUCT

The new San Diego Aqueduct would traverse some of the most rugged and inaccessible terrain in the inland area of San Diego County. Wartime priorities

San Diego Aqueduct under construction
San Diego County Water Authority

in the allocation of needed materials such as steel and cement and shortages of
experienced construction labor severely complicated the project. The route included seven tunnels, all in San Diego County, ranging in length from 500 feet
to 5850 feet. Except for almost 2 miles of steel pipe, the aqueduct used rein-
forced concrete pipe. The tunnels were concrete-lined. No pumps were required

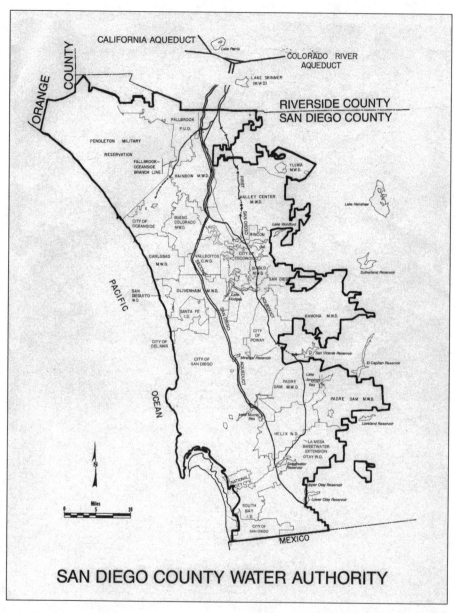

San Diego Aqueduct   *San Diego County Water Authority*

to move the water since it was gravity flow all the way. Terminology about this water-moving arrangement can be confusing; the word "aqueduct" is some-times used to describe the pipe through which the water flows; technically, though, the aqueduct is the entire system including the tunnels and can consist

of as many pipes as the aqueduct will hold. The pipes, by the way, are sometimes called "barrels" by the engineers.

The San Diego Aqueduct was placed in service in November 1947 and not
a moment too soon. San Diego was literally on the verge of running out of
water; rationing was imminent. When the aqueduct was put into operation, San
Diego County had less than three weeks' water supply remaining. SDCWA
attorney Bill Jennings called it a "real cliff-hanger." Fred Heilbron, who had
worked so hard to bring this about, went home in exultation. Everyone involved heaved a huge sigh of relief.

BRINGING THE FEATHER RIVER WATER SOUTH

Not many years went by before it became apparent that the Colorado River
Aqueduct would not solve all of southern California's water problems. Nor
would it solve those of central California where agriculture development took
vast amounts of irrigation water. California's overall water problems and the
prospect of shortages were made abundantly clear in the comprehensive California Water Plan issued by the California Department of Water Resources in
1957. For the first time, the people of California were given an in-depth, realistic picture of the state's water needs and resources.

The California Water Plan's headline-making proposal was the movement
of Feather River water down the length of California to the central and southern part of the state. Intensely controversial, it was immediately perceived as
pitting water-plentiful north against water-starved south.

The Feather River Project was promoted as bringing an eventual 2.5 million acre-feet of water each year to southern California water agencies, almost
double the amount they had been receiving from the Colorado River Aqueduct
(that goal, by the way, has never been achieved). The California Water Plan and
the Feather River Project were fought over in succeeding sessions of the state
legislature for almost a decade from 1951 to 1959 and the emotional north-
south division came close to killing any hopes for the proposed statewide plan.
Finally, in 1960 the plan was submitted to the voters together with a proposed
bond issue to provide $1.75 billion for the project that would bring the Feather
River water south.

A vociferous and divisive statewide battle ensued. Those in favor argued
that about 70% of total stream flow occurs north of the latitude of Sacramento
but over 75% of the state's water usage is south of that line. And that northern
California had two-thirds of the water while southern California had two-thirds
of the population. Speakers for the northern California opposition to the project
sometimes cited the Owens Valley episode, crying "don't let Los Angeles take
our water like they raped Owens Valley."

Surprisingly, southern California was not united in support of the new plan. The City of Los Angeles and the influential *Los Angeles Times* criticized the proposition. San Diego, on the other hand, went all out. Leading the fray were Fred Heilbron, chairman of SDCWA and Bill Jennings, its general counsel. San Diego could take great credit for the plan's approval when the measure was carried by about a 200,000-vote statewide majority, precisely the margin of approval in San Diego County. Los Angeles not only opposed the state plan but also was a negative factor in the ensuing fight to authorize the new California Aqueduct intended to bring the Feather River water south.

The result of adoption of the new plan was a new aqueduct running down the center of California; it was called the California Aqueduct. The starting place for the new aqueduct was Oroville, where the nation's highest dam, Oroville Dam (at 770 feet), was built astride the Feather River to form Lake Oroville which stores water for the State Water Project. When needed, water is released from Lake Oroville into the Feather River which flows into the Sacramento River. The water then flows into the north end of the huge delta formed where the Sacramento and San Joaquin rivers meet northern San Francisco Bay.

Originally, it was hoped by the planners that a pipeline could carry the Feather River water across the delta but this solution evaporated in controversy and concern over its high cost. Next, consideration turned to the Peripheral Canal option that would carry the water for 42 miles (in an unlined ditch 400 feet to 500 feet wide) skirting the eastern edge of the delta. The legislature approved this approach and placed the necessary bond issue on the ballot. It was defeated in 1982, with 73% of the voters in San Diego approving and 62% voting against statewide.

Water flow through the delta was left as the only available solution. Most of that water comes from the joining of the Feather and Sacramento rivers, the balance is from upstream reservoirs. Water from south delta is lifted by huge pumps into the 444-mile California Aqueduct, an open canal running by gravity flow down central California to the point where it "leapfrogs" about 2000 feet over the Tehachapi Mountains and flows down into water-thirsty southern California. This was a spectacular engineering feat, lifting huge amounts of water over perhaps the highest distance in history in an area haunted by two major earthquake faults.

THE SECOND FATEFUL DECISION

Los Angeles and MWD had no interest in seeing any of the Feather River water go directly to San Diego, but San Diego itself could have opted to become a direct contractor with the State Water Project by making a direct connection with the new California Aqueduct, thereby bypassing MWD entirely. In fact, this was given serious consideration as far back as 1955, when the

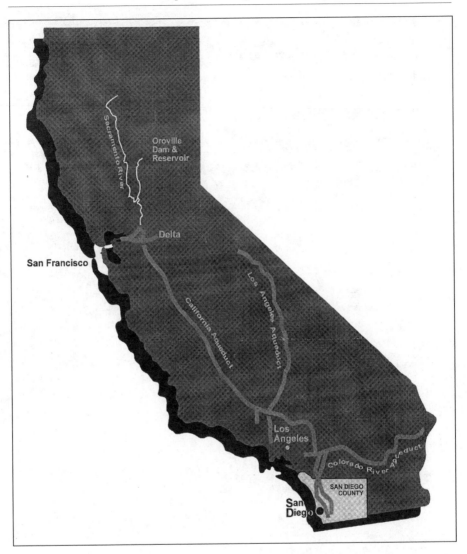

California aqueducts   *San Diego County Water Authority*

State Water Project was in its early study stage. However, when decision time came, San Diego was preoccupied with plans to build a new pipeline to bring more Colorado River water south from Lake Skinner in Riverside County down into San Diego County and wanted MWD to share in the cost.

So, in a decision that conceptually was not much different from the years-earlier choice to take Colorado River water from the MWD system rather than building an aqueduct west to join with the All-American Canal, San Diego chose once again to rely on MWD for imported water, this time for the Feather River water. Instead of proceeding independently, San Diego used the "threat"

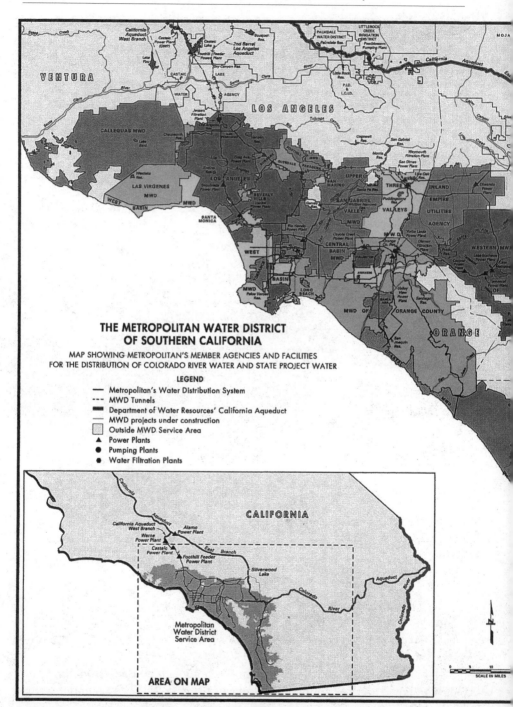

California Aqueduct East Branch

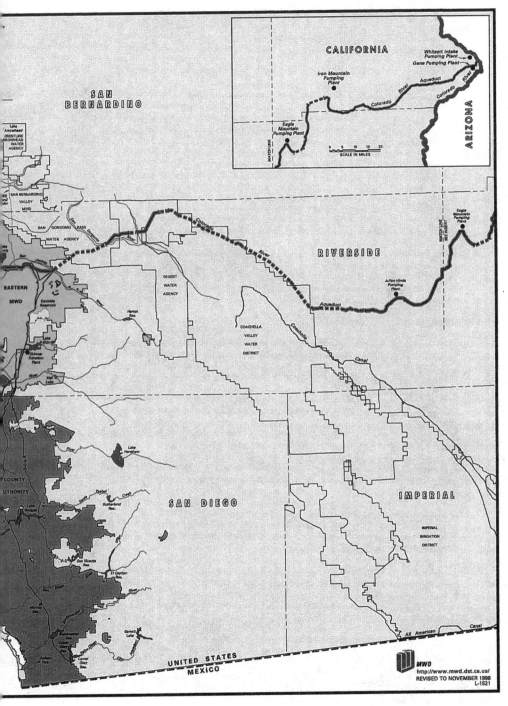

Metropolitan Water District

of hooking up directly for the Feather River water to help force MWD to make the desired cost contribution for its new pipeline. MWD agreed and provided the money for the pipeline.

For the sake of those funds, San Diego and its countywide agency, SDCWA, lost another opportunity to gain a measure of water-supply independence from MWD. And this time, the decision was more costly since the Feather River water itself would have been free to San Diego. Now, San Diego would have to pay MWD for all of its imported water. Water attorney Bill Jennings recorded that this was a "dangerous decision" by San Diego.

Thus, two fateful decisions were made over the years that set the course for the future of SDCWA's use of imported water. The decisions confirmed total control by MWD over San Diego's access to water from both the Colorado River and the Feather River, a control that has rankled San Diego as the years went by.

CHAPTER 6
A CONSTANTLY TROUBLED MARRIAGE

Distribution of the imported water that comes to southern California from the Colorado and Feather rivers is by state law the responsibility of two major umbrella agencies, MWD and SDCWA, with the latter a member agency of the former. For years, the two agencies have found reasons to disagree over important water issues.

THE TWO GOVERNING WATER AGENCIES

MWD came first. Created by the state in 1928, it was given immediate responsibility for building both Parker Dam and the Colorado River Aqueduct and its execution of those formidable tasks has been previously described. At first, there were 13 cities in MWD and although it was contemplated from the beginning that San Diego could receive the water, it was not one of the 13. Today, MWD delivers imported water from the Colorado River and Feather River through its 26 member public agencies to 18 million people throughout southern California.

The MWD board of directors that makes policy is composed of representatives from all of its member agencies; San Diego supplies 4 of its 37 members. Allocation of voting power is based by state law on a property tax formula; members with higher amounts of assessed property tax values have the most votes. The City of Los Angeles, with its highest property tax assessments, obviously had the most, although less than 50% from the beginning. It has always been an MWD anomaly that Los Angeles with the greatest voting power has the least need for imported Colorado River and Feather River water since the Owens River has over the years supplied most of its needs.

As noted earlier, SDCWA did not come into existence until the water from the Colorado River Aqueduct reached southern California in 1945 and the Secretary of the Interior needed a specific agency to receive San Diego's share of the water (the records were ambiguous as to whether the original filing was by the city or the county). So, in Sacramento, SDCWA was formed by state law. Known casually as the Authority, it distributes imported water throughout San Diego County through its member agencies. As local supplies have diminished relatively over the years, most of the county's needs have been supplied by

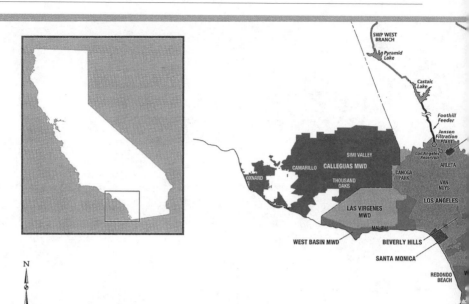

N

METROPOLITAN'S MEMBER AGENCIES

LEGEND

— Department of Water Resources' California Aqueduct
— Metropolitan's Colorado River Aqueduct
◉ Water Filtration Plants

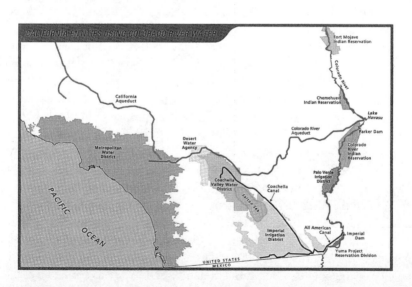

MWD member agencies

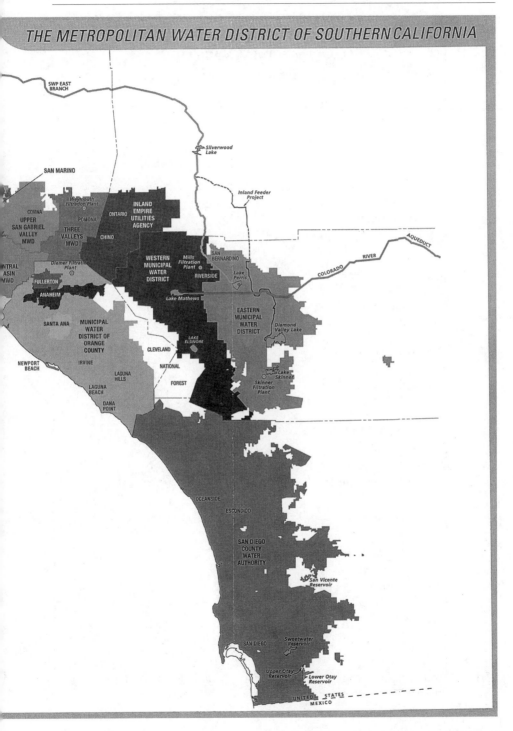

Metropolitan Water District

METROPOLITAN WATER DISTRICT
VOTE ENTITLEMENT OF MEMBER PUBLIC AGENCIES
AUGUST 19, 2003

Member Agency	Percent of Total Vote
Anaheim	1.84
Beverly Hills	0.98
Burbank	0.90
Calleguas MWD	4.04
Central Basin MWD	5.71
Compton	0.16
Eastern MWD	2.20
Foothill MWD	0.63
Fullerton	0.73
Glendale	1.14
Inland Empire Agency	3.63
Las Virgenes MWD	0.95
Long Beach	1.80
Los Angeles	19.40
MWD of Orange County	17.21
Pasadena	0.96
San Diego County	17.66
San Fernando	0.08
San Marino	0.21
Santa Ana	1.09
Santa Monica	1.17
Three Valleys MWD	2.57
Torrance	1.18
Upper San Gabriel Valley MWD	3.66
West Basin MWD	6.80
Western MWD	3.32

Note: Total vote entitlement is 134,282. Each agency has one vote for each $10 million of assessed valuation of property taxable for district purposes.

imported water. As with MWD, SDCWA policies are set by its board of directors, composed of representatives from the water agencies it serves.

The governing boards of both MWD and SDCWA are almost completely insulated from public accountability. The only link to the public is through the board members, who are appointed from cities and water agencies that are governed by elected officials. In practical operation, while both boards have chairpersons, both MWD and SDCWA are today run by chief executives chosen by their boards of directors. The current chief executive at MWD uses the title "president and CEO"; at SDCWA, that position is called general manager.

EARLY STRAINS ON THE MWD-SDCWA RELATIONSHIP

When the deal was finalized to bring San Diego's share of the Colorado River water via the Colorado River Aqueduct built by MWD, the price paid by San Diego included annexation of SDCWA by MWD. Thus, the relationship between the two agencies had from the beginning elements of a "shotgun marriage," and it soon became clear that, to use the phrase of water authority Philip Pryde, it was an "unhappy marriage where divorce is impossible."

In Sacramento at the beginning of MWD's operations, the agency was often referred to simply as "Los Angeles" even though that city had far less than a majority vote on MWD's board; this perception has clouded the relationship between the two agencies from the beginning. The emotional differences between San Diego and Los Angeles have always been present and Bill Jennings, the longtime attorney for SDCWA, opined that "this feud will continue for years until this whole southern California area is pretty well one homogenous area." Which it has not yet become and there are few signs that it ever will.

Another basic reason for the rocky relationship has been the perception that Los Angeles needs the imported Colorado River and Feather River water much less than San Diego. San Diego was desperate for water at the end of World War II; it absolutely could not survive any longer on the three "R's": rivers, rainfall, and reservoirs. The City of Los Angeles could and still does in great part, thanks to that man named Mulholland, who gave that city a direct pipeline to the water-rich Owens River. It has seemed for many years that there would be no end to the waterly beneficence of the Owens River system. Besides, the Owens River water is markedly cheaper and of better quality than Colorado River and Feather River water.

Another complicating factor is that MWD's voting and water distribution preferences are both based on the property tax bases of the member agencies. The City of Los Angeles, for years a very small user of imported water because of its Owens River supply, wielded stronger power at MWD because of its far larger property tax base. San Diego, on the other hand, although it is MWD's largest customer, was relegated to a much lower power position because its property tax base is much smaller than that of Los Angeles. Quite obviously, since the adoption of California's Proposition 13 that had a marked negative effect on property tax flows to local governmental agencies, MWD's revenues from property taxes of member agencies have decreased while as imported water consumption grew and the prices for that water were raised, its income from sales of water have greatly increased.

As MWD has become over the years more dependent on San Diego to cover its capital costs and expenses, San Diego did not gain any additional leverage within MWD. Los Angeles, on the other hand, paid less to MWD for water but maintained its power position. In recent years, Los Angeles has in some years increased its purchases from MWD but the basic imbalance remains. With respect to its voting power on the MWD board, Los Angeles can regularly pick up votes from smaller, nearby water agencies. San Diego purchases more water but still has far less voting power than does Los Angeles and its allies. And that situation continues to gall San Diego.

PIPELINE DISPUTES OF THE 1950s

The original San Diego Aqueduct (with its single pipeline) did not long satisfy San Diego's soaring post-war water needs. As previously noted, it was designed to fill only the Navy's projected water needs in San Diego County. For this reason, the pipeline was about half the size that would eventually be needed and that day arrived early in the 1950s, much sooner than anyone had anticipated. Fortunately, the Reclamation Bureau which built the pipeline, while forced to accept the Navy's decision regarding the size of the pipe, had acquired sufficient right-of-way and built large enough tunnels and siphons to accommodate another, parallel pipeline.

San Diego could not use a bond issue to build the second pipeline since it did not own the aqueduct that would carry the pipeline; it was leased from the U.S. Reclamation Bureau. The MWD board was, according to attorney Bill Jennings, lukewarm if not antagonistic to the project. He also recorded that the MWD attitude angered Fred Heilbron, chairman of SDCWA: "Them fellers up in Los Angeles just don't like San Diego. They forget that we brought with us 112,000 acre-feet of water when we joined and we intend to get it." This was the beginning of a long series of disputes that some called a "running feud" between Heilbron and Joe Jensen, the Los Angeles chairman of MWD. Harry Griffen, who knew them both well, described them colorfully in his *Harry Griffen Manuscript* as "strong-minded, opinionated, sincere, resourceful and loyal to the cities they represented."

The Navy, while seemingly cooperative, was still not formally committed to the new pipeline sought by San Diego. Finally, Secretary of the Navy Dan Kimball was persuaded to come to Los Angeles for a meeting on the matter.

Bill Jennings tells the story. When Jensen of MWD learned that Kimball was going to be in Los Angeles, he invited the Secretary to a 7 a.m. breakfast meeting with MWD members at his club, informing Heilbron in San Diego by telephone at the last minute; almost as an afterthought it seemed. Heilbron, sensing an end run, immediately corralled the SDCWA attorney Bill Jennings and, leaving San Diego at 5 a.m., they arrived at the club in the nick of time. Jensen was informally chairing the meeting and, recalled Jennings, "It was obvious to Fred and me that the thing was pretty well rigged to present the MWD view and not the position of San Diego." So, Jennings decided that it was "now or never" and before Jensen could start speaking as the breakfast came to a close, he "butted right in without giving Jensen an opportunity to speak: 'Excuse me, Mr. Secretary, I would like to say a few words.'" Without further ado, Jennings launched into a lengthy, eloquent, and fact-filled description of San Diego's need for federal governmental funding of the pipeline.

When he finished, there was silence around the table. Then, Secretary Kimball turned to his host: "Well now, Mr. Jensen, I presume that Metropolitan recognizes this situation and is willing to go along with it." Jensen squirmed and dodged: "Well, we haven't come to a conclusion; we haven't agreed." Impatient, Kimball retorted quickly, "It looks to me like it is a good proposition and the Navy can endorse it. Now I've got other business to attend to and I appreciate the breakfast."

Kimball stood up and walked out on the startled and silent men. MWD's Jensen turned to Jennings and Heilbron and said grimly, "You apparently won this round, but look out for me the next time we disagree." The "next time" would be in the dispute over the Feather River water East Branch pipeline when MWD decided to strike back at San Diego.

Kimball was as good as his word. The Navy was instructed to back up the San Diego position and the Bureau of Reclamation was cooperative. Senators Knowland and Nixon and Representative Clinton McKinnon got the necessary bill through the Congress with Fred Heilbron speaking at the key committee meetings. Jennings recorded that he "did a real swell job selling the Congressmen."

PREFERENTIAL RIGHTS AND WATER PRICING DISPUTES

Preferential rights to Colorado River water at MWD, or priorities of distribution in times of water shortage are, like voting powers, based on a formula contained in Section 135 of the agency's founding statute. That formula, like the one governing voting power, was consistently interpreted by MWD to be based on property tax assessments and therefore property tax payments to MWD, not amounts paid to MWD by its member agencies for the purchase of water. Once again, as with voting power, the formula obviously favored Los Angeles since it had by far the highest assessed property tax base of all member agencies, including San Diego. San Diego has constantly complained that it consistently buys a high percentage of MWD's imported water but because of Section 135's formula, dated to 1931, it has a "preferential right" to a far smaller percentage of MWD's water.

As the years went by, property tax revenues declined, in great part because of Proposition 13, and revenues from the sale of water largely took up the burden of supporting MWD. Thus, Los Angeles continued to maintain its preferential position under Section 135 based on property tax assessments, even though it became, relatively, a much smaller financial supporter of MWD in comparison to San Diego. Over the years, SDCWA members and staff gave considerable thought to going to the legislature in an attempt to amend Section 135 and substitute a more equitable formula that would expressly give consideration to all sources of MWD income, including amounts paid for water purchases, in determining both water preferential rights and voting rights. MWD did finally agree to cosponsor a bill changing the statutory formula but the bill died in the legislature.

There were other member agencies of MWD concerned about the preferential rights situation, fearing that water would not be available to meet their needs in a time of shortage. This was particularly true with respect to newly developed areas that wanted to obtain imported water but were worried that MWD would not satisfy their needs. This and Fred Heilbron's constantly raised voice about San Diego's lack of preferential standing for water supply in times of shortage led to a special meeting of the MWD board at Laguna Beach where a declaration was adopted (later called the Laguna Declaration), stating that MWD recognized and accepted the responsibility of meeting the needs of all member agencies in a time of water shortage.

Bill Jennings, San Diego water attorney, recorded that on the preferential rights issue, "We appeared before them [MWD board] many times.... But the Metropolitan people stood very firm and kept together as a unit and they have prevailed." In 1961, though, Los Angeles wavered. A Memorandum of Understanding was signed that provided a change in the preferential rights formula to better reflect the amounts paid to MWD through imported water purchases by San Diego. However, Los Angeles soon backed away and never joined San Diego in seeking implementation of the memorandum.

The continually roiling controversy also involved pricing for the imported water by MWD. Fred Heilbron, the strong-voiced chair of SDCWA, complained constantly in those early years about the repeated efforts of the City of Los Angeles to shift MWD costs to revenue from sales of water, a program that affected San Diego more than any other member agency of MWD since it bought the most water. He and others have charged that representatives of the City of Los Angeles on the MWD board have, over the years, led the way in raising the prices paid by member agencies for imported water, creating a structure that more and more transferred MWD's costs onto the large water users such as San Diego.

SDCWA attorney Bill Jennings described those years, stating that at the time, "the ability of the MWD directors to take steps against the [San Diego] Water Authority in punishment for having kicked over the prices is considerable. But we have a very minor position in the vote on the [MWD] Board of Directors and so we're dependent, to a considerable extent, on the good will and sense of equity of the Metropolitan directors towards this area, which is not as homogeneous to the Metropolitan control areas as Los Angeles and Orange counties and its various other members."

Harry Griffen decried this controversy between SDCWA and MWD as being too often based on the emotional antipathy that San Diego has always had for Los Angeles. Not so, says a highly placed official at SDCWA who prefers to remain anonymous: "There is nothing emotional about it. A cold-blooded analysis of the undeniable facts in those years reveals a constant effort by Los Angeles to keep San Diego under the thumb of MWD insofar as water supply is concerned, to shift the costs of MWD to its customers including mostly San Diego and to preserve a dominant position for the City of Los Angeles on the MWD board."

DIFFERENCES OVER DELIVERY
OF FEATHER RIVER WATER

While some have argued that San Diego had exercised bad judgment in not seeking and paying for a direct link to obtain Feather River water from the new California Aqueduct that would bypass MWD completely, it woke up when state planners turned to implementing the route the Feather River water would take south from the end of the California Aqueduct after it crossed the Tehachapi Mountains north of Los Angeles.

MWD wanted one pipeline running directly south to Los Angeles, arguing that San Diego did not need the Feather River water and should be content with the water it received from the Colorado River through MWD. SDCWA disagreed, arguing for the so-called East Branch that would run southeasterly from the end of the California Aqueduct to a point where SDCWA could pick up the water from both Colorado River and Feather River. While MWD still controlled both sources of supply, at least SDCWA could get water from both when needed.

But this was MWD Chairman Jensen's opportunity for a "payback time" for his earlier humiliation by San Diego's representatives (Heilbron and Jennings) in front of Secretary of the Navy Dan Kimball. According to Harry Griffen, this was an easy call for Jensen: "Since the Authority [SDCWA] was served exclusively by the MWD, the [California] aqueduct need not run that far." In other words, MWD could argue "we've got San Diego over the barrel in supplying all their Colorado River water so why let them have this branch?" Acting on the MWD position, the State Engineer killed the so-called East Branch. "Once again, an independent source of water supply for San Diego was thwarted by others," Harry Griffen noted in his *Harry Griffen Manuscript.*

Fortunately for San Diego, the state engineer's decision did not end the matter. The dispute went up to the governor's office. There, after hearing the details, Pat Brown ruled that the East Branch should be built. As a result, San Diego did finally get access to the Feather River water, but instead of getting it directly and without payment for the water, as it might have done earlier, it was left with buying that water as well as the Colorado River water from MWD.

This victory still did not give San Diego untrammeled access to Feather River water. That water was still mixed by MWD with Colorado River water and before it came to San Diego was treated at the MWD plant on Lake Skinner. To this day, there are those in San Diego who argue that SDCWA made a serious mistake in not insisting on building its own water treatment plant for all imported water. Again, San Diego made a decision to accept MWD power over its imported water and again, MWD could accurately claim that it subsidized capital costs of San Diego's imported water supply.

SAN DIEGO'S "WAKE-UP CALL"

The disastrous drought of 1987-1991 created considerable tension on the MWD board. The City of Los Angeles was largely unaffected by the drought as the Owens River water continually flowed into the city mains. However, in San Diego where the reservoirs were rapidly emptying, the City and County were forced to take severe measures to reduce water consumption, much tougher than those adopted in any previous dry period. San Diegans got the message, at least temporarily, that water should not be taken for granted.

The MWD board voted to impose a 50% reduction in all deliveries of imported water. This would have been devastating to SDCWA's residents and businesses and only a very heavy "miracle March" rainfall in 1991 stopped this drastic cutback. It is said that man-made prompting played a part in that dramatic, drought-ending rainfall. Cloud-seeding by airplanes was paid for by the San Diego Water Department in an attempt to end the deadly drought and, shades of rainmaker Charles Hatfield's mysterious works back in 1916, the deluge came soon after the planes had landed. Asked if the cloud-seeding had caused the deliverance, the San Diego Water Department spokesman responded with a straight face, "We don't have any data on that and I don't know if we ever will."

The MWD board finally adopted a 20% cutback to be applied across the board to all its customers. By definition, such a cut is not based on need and could have hurt a large user like San Diego much more than one like Los Angeles that had far less need for the Colorado River water. Mike Madigan, former SDCWA Chair, states that the City of Los Angeles went largely unscathed by MWD's reductions in supply and made only "patty cake cuts" in water conservation. On the other hand, San Diego was hardest hit by the reductions. After MWD refused to obtain additional supplies from the State Water Authority, SDCWA had to purchase State Water Bank supplies at an excessive cost in order to keep adequate water flowing to San Diego County.

San Diego felt betrayed. According to the Laguna Declaration, need would dictate the allocation of water to all member agencies. Quite clearly, Los Angeles had very little need for the imported water whereas San Diego needed it desperately. Fortunately, that "miracle March" rainfall ended the dispute. However, both the drought and MWD's initial reaction to it constituted a very loud wake-up call to San Diego. Madigan concluded, "This crystallized the conclusion by many of us at SDCWA in those years that San Diego could not depend on MWD in a water crisis situation." And many in San Diego felt that the events showed Los Angeles' control over MWD. Madigan said it succinctly: "When Los Angeles wanted to control that Board, it could. Period."

An examination of the record shows that MWD's drought action plan as adopted in November, 1990, allocated water among all members according to the amount of water each had taken during 1989-90. For example, for the 1989-1993 period, San Diego's share was .537 million acre-feet (25%) and Los Angeles' share was .321 million acre-feet (15%). Preferential rights under Section 135 were not used in allocating the available supplies, but the MWD board had the discretion to change this whenever it wished.

Some at SDCWA felt that this marked a turning point in the relationship between SDCWA and MWD. Thereafter, they say, San Diego became dedicated to finding ways to lessen its dependence on MWD, a seeking for independence, some called it.

MWD-SDCWA POLICY DIFFERENCES IN THE 1990s

As the years went by, MWD and SDCWA developed divergent views on some critical water supply issues. Two strong representatives from San Diego on the MWD board, Harry Griffen and Hans Doe, had taken the lead during the 1980s in fostering what came to be called a "regional approach" to water supply issues. These problems affect all of southern California, they reasoned. So, since we're all in this together, let's work together seeking to find both consensus and solutions.

Lin Burzell, then General Manager of SDCWA, recalls vividly the time when Heilbron came onto the MWD scene. As previously mentioned, Heilbron was a tall, commanding figure who tended to dominate meetings. Bill Jennings recalled that "Heilbron conducted Authority affairs with an iron hand." He had been a San Diego City councilman for years and when he was chosen to represent SDCWA on the MWD board, he moved rapidly into its leadership group, participating forcefully in policy discussions. He and Griffen, agreeing on the regional approach, frequently spoke out against what they called a "knee-jerk hostility" to Los Angeles proposals sometimes displayed by San Diego's other representatives on the MWD board and their talk about a need for more SDCWA independence on water sources.

Burzell and Griffen frequently worked in collaboration with Hans Doe, a powerful chairman of SDCWA who also served on the MWD board and was active constantly on water issues until he retired in 1986. Although he fought Los Angeles over the East Branch along with Heilbron and Griffen, Doe constantly urged SDCWA to rely on MWD on water supply matters.

Griffen argued that the sentiment against Los Angeles was just plain "ridiculous." He wrote in his *Harry Griffen Manuscript*:

Both of the cities started out in the old Spanish days with San Diego having a small head start, but they were both about equal. Although Los Angeles grew faster, San Diego always felt it was a much better city than Los Angeles and should remain free of any domination. San Diego Bay was a much better port, although Los Angeles made a port that became much more important commercially. In many ways, Los Angeles appeared to outsmart San Diego and, as a result, people in San Diego were very jealous of their local importance and were hurt about the idea of having to join something Los Angeles had developed.

Unquestionably, Heilbron, Griffen and Doe led the way in causing MWD to give more consideration to the southern counties. Old-timer Paul Engstrand, for years counsel for SDCWA, Lin Burzell, SDCWA general manager for many years, and officials of San Diego's North County water agencies consistently supported what Griffen called the "consensus-seeking approach." They were not comfortable with arguments stressing a need for what some called "water independence" for San Diego, urging that it would be largely meaningless and could become counterproductive in the long run, given the unassailable facts that MWD has statutory responsibility over southern California's largest imported water supply from its Colorado River Aqueduct and that San Diego desperately needs that Colorado River water from MWD.

Paul Engstrand, among others, pointed to "dire consequences" if an independent posture had been followed in the past. For example, that posture could have argued against joining with MWD to obtain increases in Colorado River water that were needed as population grew. And Engstrand argued persuasively in favor of the benefits to be derived from reaching consensus agreements on the MWD board, with the two large cities of Los Angeles and San Diego coming into accord in dealing with water supply issues in a way that could benefit everyone in both cities.

In a recent interview, Engstrand called attention to the acts of forbearance by MWD in the past such as never invoking the controversial Section 135 to San Diego's disadvantage. He also stressed MWD's contributions to the infrastructure needed by San Diego to obtain its Colorado River water, mentioning both the financial support for the Lake Skinner plant that treats the water purchased by SDCWA from MWD, and the East Branch aqueduct that brought Feather River water to San Diego. He stated that the voting structure at the MWD board always permitted decisions "representing consensus fairly achieved."

As the end of the century approached, it became clear that those who favored the so-called "regional approach" did not control the SDCWA board. It has been suggested that San Diego's posture changed when Mayor Pete Wilson appointed some very strong, issue-oriented members to the SDCWA board who went on to also serve on the MWD board, a practice continued by Mayor Murphy. Francesca Krauel and Christine Frahm, two City of San Diego representatives on the MWD board in the 1990s, certainly did not adopt overall the "regional approach" and became a rather vocal minority on the MWD board, creating what some called a "rift" within that body.

Lin Burzell recalled critically that Frahm and Krauel "persisted in running down major projects of MWD, seeming to seek ways to say that MWD was wrong." An outsider, Professor Steven Erie of the University of California at San Diego, who has written frequently on water issues, states that Frahm particularly was viewed as the "Rodney Dangerfield" of the MWD board. The Frahm-Krauel activities are said by some to have been another precursor of the later, stronger movement toward SDCWA taking a more independent posture with respect to MWD.

Another strong member of the SDCWA board representing the City of San Diego during the 1980s and 1990s was Mike Madigan who for a time served as chair of the board. Mike has consistently decried the regional approach and regretted the power that the City of Los Angeles has often exercised over MWD policies. When reminded that his words sound like a personal echo of the age-old hostile San Diego-Los Angeles refrain, Madigan responds firmly and revealingly, "Not me. I like Los Angeles but I happen to live in San Diego."

CHAPTER 7
WATER DISTRIBUTION
IN SAN DIEGO COUNTY

The umbrella water agency previously discussed, San Diego County Water Authority, supplies virtually all of San Diego County water needs for nearly 3 million residents and a $126 billion economy. However, distribution to meet those needs covers geographically only about a third of the county, where practically all the water users are located. The balance of the county, almost entirely mountain and desert areas, depends on rainfall and underground water.

SDCWA reports to no elected official in San Diego County, although each of its member agencies does have city council or water district officials who are elected. Also, the City of San Diego has representatives on the SDCWA board who are appointed by the elected mayor. Established by state statute, SDCWA has a daunting mission to supply a safe and reliable source of water to all users in San Diego County, one of the most populous counties in the nation. Its power is limited to the imported water it sells to member agencies throughout the county, but to call it simply a "water wholesaler" would be very misleading. True, its major function involves the selling of water but its overall role is considerably greater. Ensuring that the county's water supply is both safe and reliable is an extremely broad and heavy responsibility.

The Authority's total system capacity is approximately 900,000 acre-feet per year. About two-thirds comes from the Colorado River and one-third from the Feather River. However, SDCWA's authority is not limited to water from those rivers. It is also authorized to acquire water and water rights anywhere within or outside California and to develop, store, sell, and deliver the water. It has powers of eminent domain and the right to levy and collect taxes and issue bonds. The bond issues approved by the voters of San Diego County have been very substantial; SDCWA bonded indebtedness went over $180 million in 2001.

OVERVIEW OF SDCWA

SDCWA currently has 23 member agencies; the City of San Diego is the largest; 5 other cities are members. Other members are 16 water districts and Camp Pendleton. The Authority is governed by a 34-member board. Each

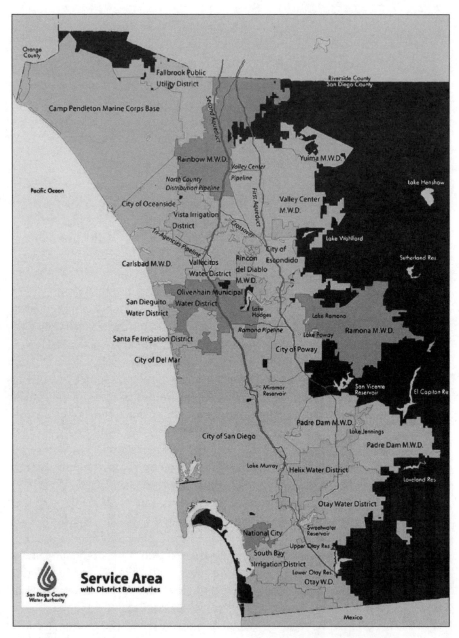

SDCWA member agencies   San Diego County Water Authority

SDCWA Member Agency Voting Entitlements
(As of January 1, 2002)

Agency	Total Financial Contribution[1]	Votes[2]	Percentage
Carlsbad Municipal Water District	$179,576,376	35.92	3.26%
City of Del Mar	17,382,560	3.48	0.32
City of Escondido	171,934,619	34.39	3.13
Fallbrook Public Utility District	144,235,337	28.85	2.62
Helix Water District	431,318,497	86.26	7.84
City of National City[3]	49,308,724	9.86	0.90
City of Oceanside	261,287,255	52.26	4.75
Olivenhain Municipal Water District	114,887,333	22.98	2.09
Otay Water District	220,187,912	44.04	4.00
Padre Dam Municipal Water District	188,204,560	37.64	3.42
Camp Pendleton Marine Corps Base	11,185,812	2.24	0.20
City of Poway	112,066,657	22.41	2.04
Rainbow Municipal Water District	240,043,029	48.01	4.36
Ramona Municipal Water District	85,556,040	17.11	1.56
Rincon del Diablo Municipal Water District	87,099,178	17.42	1.58
City of San Diego	2,310,040,328	462.01	41.99
San Dieguito Water District	63,951,066	12.79	1.16
Santa Fe Irrigation District	80,824,641	16.16	1.47
South Bay Irrigation District[3]	149,823,374	29.96	2.72
Vallecitos Water District	101,617,229	20.32	1.85
Valley Center Municipal Water District	311,680,069	62.34	5.67
Vista Irrigation District	150,507,772	30.10	2.74
Yuima Municipal Water District	18,201,065	3.64	0.33
TOTAL	$5,500,919,433	1,100.19	100.00%

[1]Total financial contribution equals base revenues plus fiscal year 2001 revenues.
[2]Rounded values.   [3]Comprise Sweetwater Authority.

SDCWA member agency voting rights   **San Diego County Water Authority** member agency has at least one representative on the board and the County of San Diego has an honorary position. Fred Heilbron chaired the board for over 30 years, as one observer judged, "with a combination of iron hand, wit, and benevolence," becoming known in the process as a "Mr. Water" of San Diego County. He also served as Secretary of MWD for 17 years. Since his retirement, the job of chair has often been rotated between representatives of San Diego and other member agencies.

Heilbron's decisive leadership has been previously documented. Not only was he the spark for San Diego perfecting its rights to Colorado River water back in the 1920s and the driving force behind creation of SDCWA in the 1940s, he raised his voice constantly about what he called its inferior voting power on the MWD board and its inferior preferential rights to imported water in times of shortage. Heilbron ran the Authority with the help of Bill Jennings, its attorney, whose wife typed the meeting minutes at night. Jennings recalled years later that Heilbron was a "very outspoken man" who "conducted the affairs of the Authority with an iron hand." Others described him in later years as becoming more "dictatorial," often running roughshod over other members of the SDCWA board. Linden Burzell's unpublished *Memoirs* contains a story describing one of his first meetings with Heilbron at the latter's plumbing business. Then 87 years old, Heilbron kept Burzell waiting while he finished lambasting his 60-year-old son for buying a set of tires without his permission.

The current general manager of the Authority is Maureen Stapleton, a former San Diego deputy city manager. As general manager, she is responsible for directing the staff and all administrative affairs of the Authority and, along with the chairman, speaks publicly for the Authority. Highly regarded by members of the board, she is often described as a "strong leader," particularly with respect to the Authority's current efforts to diversify SDCWA's sources of supply for water.

The City of San Diego is limited by the State Water Authority Act to a 50% weighted vote on the agency's board of directors. However, on most matters coming to a vote by the board, the voting was traditionally conducted on a member basis with a majority controlling. Each member agency votes on a unit basis, so San Diego has one vote as does each other member agency; those with more than one representative must decide how to cast their single votes. But on any issue that it chooses, any member agency can demand a weighted vote. In this situation, San Diego has consistently had the largest vote since property taxes paid were originally given the heaviest weight in the formula used to measure weighted votes.

Traditionally, the City, conscious of the fact that county agriculture has importance far beyond its numbers of people and property tax assessments, has consistently refrained from demanding a weighted vote on any issue involving agricultural water concerns. Finally, in 1996 repeated disputes arose over the voting formula and the board called in Mike Madigan (he had earlier left the board after serving for 15 years but returned to help resolve the disputes). Some wanted "one man, one vote" for each agency served by SDCWA. The solution was a new voting system based on water usage; those who used the most water had an increased "weighted" vote. Not coincidentally, it is basically the same system that SDCWA has been for years fighting for at MWD.

SDCWA serves water users scattered over 909,000 acres in San Diego County and population in those acres pushes 3 million. The water served out by SDCWA goes through 274 miles of pipeline and currently there are five of these covering the county. In the fiscal year 2001-2002, San Diego County used a total of 721,652 acre-feet of water and 650,695 or 90% of that total was imported water, by far the most from the Colorado River. Looking to the future, water demand is expected to be well over 800,000 acre-feet annually by the year 2020.

SDCWA and its member agencies provide water to the County's users through a very complex system. Surprisingly, only about 17% of it goes to agricultural usage. Overwhelmingly, the major users are consumer, industrial, commercial, governmental, and military. The City of San Diego buys 37.9% of the water supplied by SDCWA. All other customers buy far less. The next largest user is Valley Center Municipal Water District at 7.8%, followed by Helix Water District at 6.4%. All the other member agencies are less than 5%.

About half of SDCWA's water is treated by MWD at Lake Skinner in Riverside County and then pumped into the SDCWA distribution system. The rest of the imported water is delivered raw to ten agencies (City of San Diego is the largest) that have local water treatment facilities necessitated by their usage of water from reservoirs and other local sources. Just a few receive all their water from neighboring agencies. All of the agencies (except Yuima that uses mostly local groundwater sources) have interconnecting pipelines with other agencies so that transfers can be accomplished in times of emergency disruption.

The SDCWA operating budget for the fiscal year 2003-2004 totals $27.5 million; that year's appropriation for the Authority's Capital Improvement Program (CIP) is $150.3 million.

NEW PIPELINES CONSTANTLY NEEDED BY SAN DIEGO

As previously discussed, difficulties were encountered by SDCWA in building the second pipeline after the first one proved to be too small. Actual construction did not begin until 1952 and SDCWA did not finally take over the pipeline for operation and maintenance until February 1955. And even the second pipeline was not enough. Like the growing boy always needing a new pair of shoes, San Diego seemed to always need a new pipeline. Another new decade, another new pipeline.

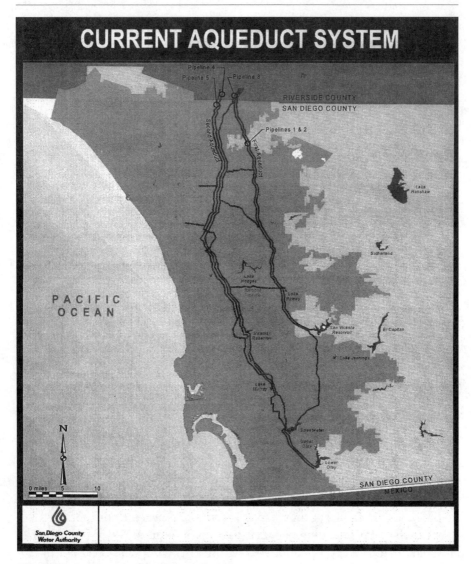

SDCWA pipelines in San Diego County
San Diego County Water Authority

Pipelines No. 1 and No. 2, each four feet in diameter, had filled up the first aqueduct, so before any more could be laid, another aqueduct had to be built. The second San Diego Aqueduct, much larger than the first, was built closer to the coast and ran from the west portal of the Mount San Jacinto tunnel terminus of the Colorado River Aqueduct down to Otay Reservoir. Installed in this aqueduct was Pipeline No. 3, with a 6-foot diameter. But the population continued to soar; as SDCWA attorney Bill Jennings said at the time, "everyone was just about half a jump behind the demands for water."

Another new decade, another new pipeline. The second aqueduct had room for it, so a new pipeline (No. 4, with an 8-foot diameter) was built, capable of carrying as much water as all its predecessors combined. Pipeline No. 4 was 33 miles long, running down to the Alvarado Water Treatment Plant near La Mesa. In the 1980s the population was pushing 2 million so Pipeline No. 5 had to be added to the second aqueduct. This raised the total pipeline capacity to about one million cubic feet/year, about 15 times the capacity of Pipeline No. 1 built 35 years earlier. Another new decade, another new pipeline. And now, with the new century just commencing, Pipeline No. 6 is being seriously considered by SDCWA.

Winding the clock back a few years, it became evident when San Diego's growth continued to escalate that SDCWA would have to change its policies. The Authority's focus from the beginning had been on areas in San Diego County where there were in-county water supply sources that needed only supplementation from the Colorado River water at San Vicente Reservoir. When the new pipelines were added, bringing to San Diego more imported water, agencies from largely rural areas with inadequate existing water supplies and wanting a "piece of the action" clamored to join SDCWA. Conservative members of the Authority's board of directors were reluctant, fearful of making water supply commitments that would be difficult to meet. The Authority decided to face up to and accept responsibility for obtaining whatever water was necessary to meet new needs of member agencies. The policy statement adopted by SDCWA declared, "When and as additional water resources are required to meet increasing needs," SDCWA "will be prepared to deliver such supplies."

RECEIVING, STORING, TREATING, AND DISTRIBUTING THE WATER

SDCWA takes water for its five north-south pipelines from the two San Diego Aqueducts at a point about 6 miles south of the Riverside County line. From there, it is distributed through 274 miles of pipeline and 119 service connections to the Authority's member agencies. The agencies that receive the water include cities, water districts, irrigation districts, one public utility district (Fallbrook), and one military reservation (Camp Pendleton).

SDCWA does not have any water treatment facilities because it buys treated water from MWD at Lake Skinner. MWD has recently decided to commence fluoridation for all the imported water it treats at Lake Skinner. The year 2006 is predicted for complete fluoridation of all treated water purchased by SDCWA and its member agencies. This step leaves open the question of fluoridation of the untreated imported water. Water agencies under SDCWA are currently considering whether to fluoridate the water they purchase untreated from MWD.

SDCWA has no responsibilities with respect to the 24 reservoirs in the county that are controlled by the agencies that own them. They have a combined capacity of approximately 571,000 acre-feet. The reservoirs are used primarily for storage, to collect the water that otherwise would simply run into the ocean, and this stored water helps San Diego survive the drought years. As will be discussed later, they would be vital in the event of an emergency disruption of the pipelines that bring imported water to the County.

The Authority is currently embarked on an ambitious Capital Improvement Program totaling $1.6 billion. Initiated in 1989, it is scheduled for completion in 2010. The program is designed to increase pipeline capacity, eliminate bottlenecks in the present system, increase reliability of the system and increase operational flexibility to facilitate pipeline maintenance. The most significant element of the program is the $827 million Emergency Storage Project designed to provide water for emergencies and disasters (see Chapter 8).

All water used in San Diego County is distributed by delivering agencies which obtain their water from local sources and/or SDCWA (the supplier of all the imported water). Six are cities: San Diego, Del Mar, Oceanside, Escondido, Poway, and National City. Sixteen are special districts: water districts, public utility districts, and irrigation districts. One, Camp Pendleton Marine Corps Base, is a federal agency. Each of the cities and districts has statutory power to issue bonds and use tax monies for operations. Each of the districts is governed by a board of directors consisting of local residents chosen by popular vote within the district's service area. Since they are separate member agencies, National City and South Bay Irrigation District (principally Chula Vista) have individual voting powers on the SDCWA board, but they are combined into the Sweetwater Authority for purposes of water distribution.

Each of the various water-distributing agencies developed historically in response to local need and available water sources. Some, like the City of San Diego, City of Escondido, City of National City, South Bay Irrigation District (Sweetwater Authority), Vista Irrigation District, and Helix Water District, have extensive local water supply sources but this supply must still be supplemented with imported water purchased from SDCWA. Others have smaller local water sources and are much more dependent on the Authority's imported water. Many have no local sources, including City of Del Mar, City of Poway, and Otay Water District, and the following municipal water districts: Carlsbad, Olivenhain, Padre Dam, Rainbow, Ramona, Rincon del Diablo, and Valley Center.

Most of the agencies having substantial quantities of local water in reservoirs, like Sweetwater for example, look at water supply on a long-term basis. Says Paula Roberts, Sweetwater spokesperson, "You have to expect that in only two years out of 10 will rainfall fill your reservoir." The reservoir water is "free" — runoff from rainfall. The imported water costs SDCWA member agencies like Sweetwater $450 to $500 per acre-foot. As of this writing, drought

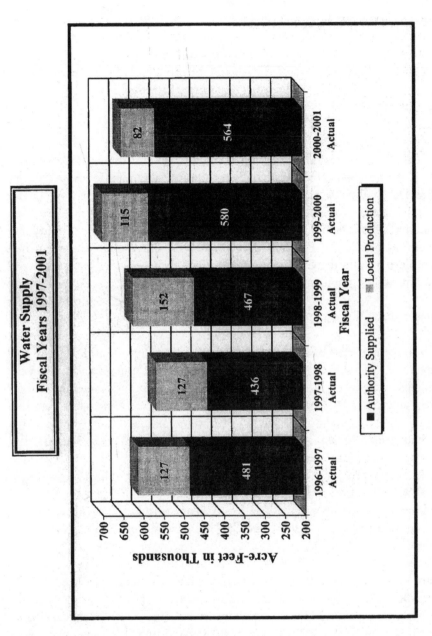

San Diego County usage of local versus imported water, 1997–2001   San Diego County Water Authority

conditions have forced Sweetwater to change its procurement balance. Usually, it is 70% local (because it is cheaper) and 30% imported; now it is the reverse and Sweetwater is planning to import 70%. The reservoir water will be "saved," available for the summer dry season and for emergency usage if service should be disrupted by some catastrophe.

The 23 agencies are divided as to water treatment. Only a few (including the City of San Diego) have their own treatment facilities. They treat imported water taken from Lake Skinner as well as water from their own reservoirs and other local sources. The other agencies purchase treated water from either SDCWA or neighboring agencies.

To provide an overview of the County's water distribution system comprising most of the County's population, this discussion will focus on the City of San Diego, four of what might be called the "suburban" districts — Helix Water District, Sweetwater Authority (including National City), Padre Dam Municipal Water District, and Vista Irrigation District — and three of the small remaining agencies — Rainbow Municipal Water District, Yuima Municipal Water District, and the City of Del Mar. Taken together, they serve well over two-thirds of the population of San Diego County.

CITY OF SAN DIEGO WATER SUPPLY SYSTEM

Following a vote of the populace in 1901, the City of San Diego (in this chapter, City) ended its concentration on private water sources and entered the municipal water supply business. One hundred years later, the City's water infrastructure is one of the most complex in the nation, serving over 1.2 million people. The City purchases all its imported water untreated from MWD, taking it directly from Lake Skinner in Riverside County and delivering it to the City's three water treatment plants via the SDCWA system.

The City is by far the largest purchaser of water from SDCWA, currently taking 38% of the Authority's total water deliveries. But not all of it is used in the City; its system also serves Del Mar, Coronado, and Imperial Beach, plus the Santa Fe Irrigation District and San Dieguito Water District. Water purchased and treated by the City is sold to the Otay Water District. Overall, City water goes through 2890 miles of water lines, helped on its way by 45 separate pumping plants. Over 200 million gallons of potable water is stored in 32 standpipes, elevated tanks, and concrete and steel reservoirs.

The City's imported water bought from SDCWA comes from Lake Skinner and accounted in 2002 for 94% of the City's water. That percentage goes down, obviously, when rainfall goes up and more water can be obtained from the City-owned reservoirs. Commonly referred to as the San Diego City Lakes, they are Morena, Barrett, El Capitan, San Vicente, Hodges, Miramar, Murray, Lower Otay, Upper Otay, and Sutherland. These lakes are actually

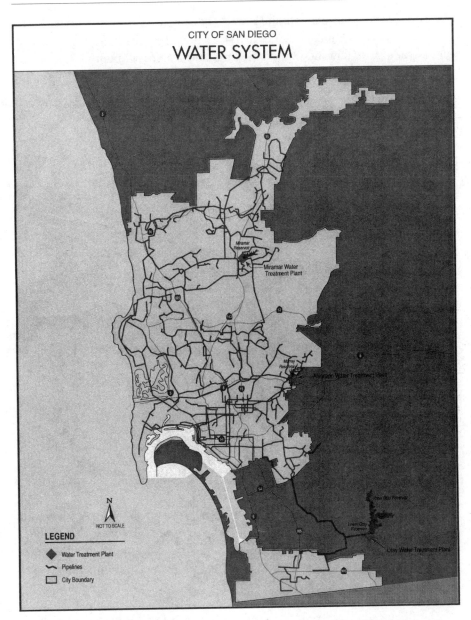

Delivery of water in City of San Diego   *City of San Diego Water Department*

storage reservoirs with some water drawn off for regular use and the balance reserved for droughts and emergencies. Present City policy requires that the reservoirs collectively maintain water sufficient for about seven months of usage, more than enough time (according to the planners) to recover from disruptions of supply and to last through severe drought periods.

To meet popular demand, the reservoirs provide recreational access for a wide variety of traditional outdoor activities including fishing, boating, canoeing, kayaking, sailing, hiking, picnicking, and waterfowl hunting plus (since 1989 at some reservoirs) seasonal programs for water-skiing, jet skiing, and windsurfing.

The City operates three water treatment plants. The Alvarado plant serving central San Diego (northern National City border to the San Diego River) takes water from El Capitan, San Vicente, Sutherland, and Lake Murray reservoirs. The Otay Water Treatment Plant serving South County (National City to the Mexican border) takes water from storage at Lower Otay, Barrett, and Morena reservoirs. The Miramar Water Treatment Plant serves North County (San Diego River to Orange County line) and draws from Miramar Reservoir. Hodges Reservoir (Lake Hodges) is not connected to a water treatment plant and sells untreated water primarily for agricultural usage to neighboring water agencies.

The accompanying map shows how the water is brought to city users from the reservoirs. By far the greatest number of service connections in the City, about 80% of the total, is single family domestic. Industrial-commercial usage accounts for about 20%.

Larry Gardner, Director of the City of San Diego Water Department, commented about what he sees is the responsibility of his department to ensure a reliable water supply for the city. Agreeing with SDCWA policy that "diversification of supply is an absolute must" and refusing to abdicate all diversification action to SDCWA, the department has prepared the City's 2002 Long-Range Water Resources Plan which emphasizes a variety of steps to improve supply through conservation, reclamation, and groundwater usage. In 1997, the Water Department established a new capital improvement program calling for over $700 million to improve, expand, and seismic retrofit all water storage, treatment, and distribution systems.

Gardner speaks readily about diversifying water supplies, discussing the negotiation of water transfers with far-away agricultural areas. Under Gardner's direction, the City is taking a hard look at the possibility of bringing in fresh water by ships. Refitting oil tankers for this purpose has proved impractical so attention has shifted to plastic bags to be towed behind ships. Natural Resources Corporation and World Water, SA, presented proposals to deliver up to 20,000 acre-feet per year of potable water to the City by the year 2004. However, the cost of building the facilities needed to move the water to the City's distribution system would be quite high, multiple permits would be required, and both governmental and public support are questionable. The City is still studying the proposals.

When questioned about planning for pipeline disruptions and other possible catastrophes, Gardner stated that the City has a detailed emergency plan but asserted that it is "secret" and "not available" for evaluation by outsiders.

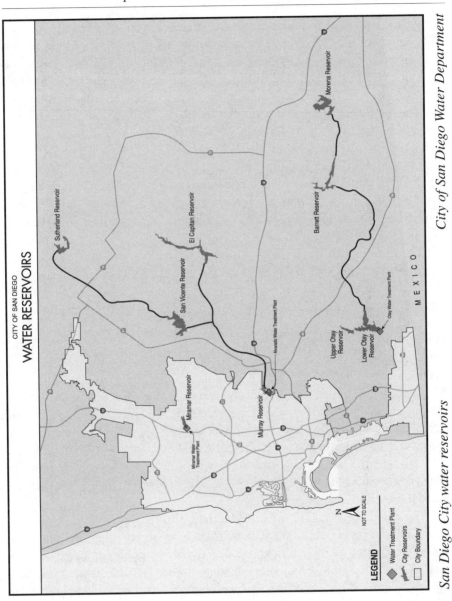

San Diego City water reservoirs

Gardner did avow that "the City Water Department is fully equipped to provide a sustained supply of water under emergency conditions for at least a six-month period of time."

THE HELIX WATER DISTRICT SUPPLY SYSTEM

The Helix Water District covers nearly 50 square miles. It includes the cities of La Mesa, El Cajon, and Lemon Grove, the community of Spring Valley, portions of Lakeside, and various unincorporated areas near El Cajon. The total population served is 251,586. Helix is an outgrowth of the old La Mesa, Lemon Grove, and Spring Valley Irrigation District, formed in 1913, that purchased the Cuyamaca Dam and the 33-mile San Diego Flume managed by Colonel Ed Fletcher. Much of the district's colorful history has been described earlier. Helix is a public agency formed under the Irrigation District Laws of California and is governed by a board of five directors elected to four-year terms, a pattern that is followed by many other SDCWA agencies.

Owning both Lake Jennings and Lake Cuyamaca, Helix also has storage rights for 10,000 acre-feet of water at El Capitan. Treated water is stored in the 30-million-gallon Grossmont Reservoir located under much of Harry Griffen Park in La Mesa and 22 storage tanks spread around the district. Total storage capacity is 63 million gallons.

The old wooden flume that historically brought water down from Lake Cuyamaca was replaced in the 1930s with concrete pipes that in turn are being replaced by steel pipelines serving about 85,000 residents in El Cajon Valley and the surrounding hillsides. It is a far cry from the old system to the 700 miles of pipe that today serve Helix District. Remains of the old flume can be seen at the Helix headquarters in La Mesa.

By far the majority of Helix water is a blend of Colorado River and Feather River water purchased from SDCWA. In the last decade, amounts from a low of 5% up to a high of 35% have been taken from local sources such as Lake Cuyamaca, Lake Jennings and El Capitan. From 1999 to 2002, extremely low rainfalls reduced the local supply to an average of 7.5%. All water goes through the district's $50 million state-of-the-art R.M. Levy Treatment Plant at Lakeside that uses the advanced ozonation process to improve taste and color. All state and federal standards for quality and safety are met and the plant's capacity is 106 million gallons per day.

The San Diego Chapter of the American Institute of Architects recently granted its coveted Orchid Award to the plant as an outstanding Environmental Solutions Project. The plant is hailed not just for its physical appearance but also for being the first in San Diego County to provide such clean, fresh-tasting, ozone-treated water that there is, as its boosters say, "no need to lug jugs of bottled water home anymore."

Helix has been a county leader in water conservation. Per capita usage in 2001 was 141 gallons per day, substantially less than the 1988 per capita usage of 164 gallons. Helix has a storage capacity of 69 million gallons of filtered water that is readily available in the event of any emergency disruption of the pipelines.

R.M. Levy Water Treatment Plant, Lakeside   *Cor Shafter Photo*

THE SWEETWATER AUTHORITY WATER SYSTEM

Today's Sweetwater Authority is another old-timer in San Diego's water story, starting out as the Kimball Brothers Company in 1869 which concentrated on serving the National City and Chula Vista areas. Those two areas were served by separate water agencies under the SDCWA umbrella, City of National City and South Bay Water District, until 1972 when they combined to form the Sweetwater Authority. A considerable amount of this Authority's water is obtained from local reservoirs and water wells.

The Sweetwater Authority serves a population of about 176,000 (including 54,000 in National City). Most (72%) of its water usage is residential, 20% is industrial and commercial, and only 8% is agricultural. The water is 30% imported and 70% local, one of the highest local water percentages in San Diego County. In contrast, imported water accounts for about 90% of usage by most of the County agencies.

Sweetwater owns and operates both Loveland Reservoir and Sweetwater Reservoir. The reservoir water is cleaned, disinfected, and treated to improve taste, color, and odor at the Robert A. Perdue Water Treatment Plant in Spring Valley, and then delivered through over 400 miles of pipelines. Groundwater is obtained by wells from what is termed the San Diego Formation, an underground aquifer. This water along with brackish (somewhat salty) water from the Sweetwater River is treated at the Richard A. Reynolds Groundwater Demineralization Facility in Chula Vista which removes salts and solids with advanced reverse osmosis membrane treatment. The modern Reynolds plant was honored in 1999 with the Orchid Award granted by the San Diego Chapter of the American Institute of Architects.

PADRE DAM MUNICIPAL WATER DISTRICT

Covering an extremely large area, Padre Dam provides service to a population of about 136,0000 on land extending from the San Diego River valley to the foothills of the Cuyamaca Mountains. Included are the City of Santee, parts of El Cajon, and unincorporated areas of Lakeside, Flinn Springs, Harbison Canyon, Blossom Valley, Alpine, Dehesa, and Crest.

Imported water accounts for most of Padre Dam usage and the balance comes from local sources. All of the water has been treated before it reaches the district. Substantial amounts of water, varying from time to time depending on seasonal demands, are kept in storage to meet emergency demands. The Padre Dam Water Recycling Facility has received international attention since 1961 and earned for Santee the title of a book, *The Town that Launders its Water*. Wastewater discharged from homes, businesses and industry is collected and processed in a multi-step facility to a level suitable for irrigation and other non-drinking purposes. Water from the facility fills the seven lakes at the Santee Lakes Regional Park and also irrigates the park landscaping. Additional millions of gallons are available for other irrigation uses in Santee. A 25-mile distribution pipeline system delivers recycled water to customer sites.

VISTA IRRIGATION DISTRICT

This district covers about 21,000 acres, over 70% of which is developed. Population is currently about 115,000. Historically, imported water purchased from SDCWA accounts for about half of the total and the balance is piped in from Lake Henshaw. During drought years, however, Vista has purchased up to 90% of its water from SDCWA. The imported water has been treated at Lake Skinner and local water is treated at Escondido's treatment plant. Vista District and the City of Escondido are still in litigation with five bands of Mission Indians over water rights to the natural flow of the San Luis Rey River.

Vista is fortunate to have a well field (24 wells) that pumps water from its groundwater basin at Lake Henshaw. During a water shortage, it could supply about 12,500 acre-feet per year. Otherwise, this water is available as an emergency reserve. Coupled with water from Lake Henshaw, there is an adequate six-month supply for use during any pipeline disruption. In addition, Vista Irrigation District has distribution connections with its neighboring water districts in Oceanside, Vallecitos, Rincon del Diablo, and Escondido so that water can be quickly transferred in time of pipeline disruption. About 80% of Vista's water usage is residential and 15% is commercial, industrial, and governmental. Less than 10% and decreasing is agricultural.

SMALLER WATER AGENCIES

The names of two of the smallest SDCWA agencies are somewhat misleading; although they are called municipal water districts, neither is a municipality. Rainbow Municipal Water District serves a population of 18,000, and Yuima Municipal Water District serves 1870 users in unincorporated agricultural areas in northern inland San Diego County. By far the most (85%) of Rainbow's water usage is agricultural and all of its water is imported, bought from SDCWA. Yuima (which includes Pauma Valley) is almost 100% agricultural in water usage and 60% of that is local water pumped from groundwater along the San Luis Rey River.

The Rainbow district straddles Interstate 15 and the San Luis Rey River and includes the small communities of Rainbow and Bonsall. The terrain is rugged and mountainous with huge groves of avocado trees dominating the hillsides. Agricultural uses are principally citrus, avocados, strawberries, tomatoes, corn, commercial nurseries, and livestock.

Fallbrook is a public utility district serving residential (primarily, community of Fallbrook), agricultural, and commercial users in North County north of Escondido and adjacent to both the Naval Weapons Station and Camp Pendleton. Population served is about 28,000 over a vast service area. The water is imported (Feather River and Colorado River) and is treated at the MWD facility at Lake Skinner. Fallbrook does add some chlorine to the water to reach an optimal range for health and safety purposes. Interestingly, the Naval Weapons Station wedged between Fallbrook and Pendleton gets imported water from the Fallbrook system. A currently pending proposal would extend imported water lines from Fallbrook into Pendleton in exchange for Fallbrook receiving storage space in the Pendleton aquifers.

In dramatic contrast with these three predominantly agricultural agencies, Del Mar, the smallest of the SDCWA agencies with a population of 4400, is almost wholly residential. Out of about 2000 service connections, almost 1900 are residential. Almost 60% of its water is residential use; over 20% is for public use largely accounted for by the Del Mar Racetrack complex. All of the water is bought from SDCWA and is treated before it reaches Del Mar.

THE REMAINING SDCWA WATER DISTRICTS

Having described some of the largest and smallest water service agencies in terms of population, and omitting Camp Pendleton Marine Corps Base (it uses local groundwater), that leaves 13 agencies. Pertinent facts on their respective populations, water supplies (imported and other), and residential-agricultural-commercial usage percentages for the year 2000 are shown in the following table.

Agency	Pop.	Imported	Other	Res.	Ag.	Comm.
Carlsbad	70,000	91%	9%	50%	25%	15%
Escondido	125,000	75	25	67	18	11
Fallbrook	30,000	95	5	43	50	7
Oceanside	161,029	93	7	61	25	10
Olivenhain	48,000	100	0	63	28	0
Otay	170,000	97	3	54	16	15
Poway	48,600	98	2	70	14	6
Ramona	40,000	99	1	47	33	7
Rincon del Diablo	27,000	100	–	67	21	13
San Dieguito	36,303	64	36	65	18	9
Santa Fe	20,635	60	40	86	8	5
Vallecitos	67,000	100	–	53	31	13
Valley Center	21,949	100	–	15	81	4

Source for data: San Diego County Water Authority

AREAS NOT SERVED BY SDCWA

All of San Diego County east of the SDCWA member agency boundaries, about two-thirds of the county's total land area, receives no imported water, no water from SDCWA. East County is a predominantly mountain and desert area, and its users are totally reliant on water obtained from rainfall and wells dipping into the county's groundwater located in underground aquifers, bodies of rock and sediment that contain water in void spaces. Many of these aquifers are found in valleys in and below the mountains, and some exist in the desert. Wells tap this water. Hydrogeologist Murray Wunderly of San Diego County's Department of Planning and Land Use states that "development in groundwater-dependent areas of the county is limited by the availability of groundwater resources." Therefore, Wunderly warns, "some areas such as Borrego Valley consume groundwater faster than it is naturally replenished. As such, water consumption at the current level in Borrego Valley cannot continue indefinitely." Borrego's serious overdraft was an issue of deep concern to John Peterson, DPLU's chief hydrogeologist for many years until his recent retirement.

CHAPTER 8
PREPARING FOR EMERGENCIES
AND DISASTERS

Concern over emergencies and disasters that could adversely affect distribution of water in San Diego County are not new. Some have been dealt with for many years; notably they include droughts, floods, earthquakes, and possible sabotage. As to the last named, the threats of sabotage made during World War I and World War II are quite similar to many of the acts of terrorism threatened today. (Editor's note: 2004 marks the fifth year of a possible "500-year drought" in the Southwest.)

DROUGHT THAT INEVITABLY WILL RE-VISIT SAN DIEGO

The droughts that hit southern California during the second half of the nineteenth century and again in the years 1987-1991 have been previously described. MWD's 1999 Water Surplus and Drought Management Plan projects that with the steps therein outlined, there is zero prospect for what the plan calls an "extreme shortage," a situation requiring allocation of available imported supplies of water. MWD has embarked on an ambitious water storage program that should suffice to meet needs during drought years.

The 1999 Plan states that "by the year 2005 with the investments made to date, Metropolitan's additional water supplies will be more than adequate to meet demands under a repeat of the 1991 drought event — even with increased demands due to growth." There is also the State Water Project's terminal reservoir storage. However, an ever-present fact of life is that imported water constitutes the major portion of MWD's water supply and that water is subject to the virtually unbridled discretion of the MWD board, a fact that has always made San Diego feel somewhat uneasy.

The State of California has the statutory right to intercede to ensure an agency such as SDCWA receives the amount of imported water absolutely necessary to meet its needs; the SDCWA 2000 Urban Water Management Plan notes this could be looked to for the protection of public health and safety. But that is a "last resort" measure and there remains San Diego's concern over MWD's application of Section 135 to lessen its preferential rights to imported water during a time of severe water shortage. This subject will be addressed subsequently.

Drought management is predicated on the almost-truism that storage during periods of surplus can equal usage during periods of drought. San Diego does have a storage supply in its reservoirs that is adequate for the normal ups and downs of water availability. But what happens if there is a severe, prolonged shortage? Unhappily, as SDCWA concedes in its 2000 Plan, "The Authority does not currently have a shortage allocation plan." However, it promises that one will become available in the near future.

CAN SAN DIEGO COPE WITH A 100-YEAR RAINFALL?

Flooding caused by heavy rainfalls has frequently brought disaster to San Diego County. Notably, the Hatfield Flood of 1916 continues to cause nightmares for city engineers. Many think this problem has been resolved by the many dams built on all of the county's major rivers. Not so. None of the dams were built with primary emphasis on flood control. True, they have mitigated water flow and the reservoirs have contained huge amounts of water that otherwise would have flooded valleys from North County to South County. For example, El Capitan Dam and, to a much lesser extent, San Vicente Dam have reduced peak flows on the San Diego River.

Not just the San Diego River, but also the San Dieguito River can still be troublesome if extremely heavy rainfall hits their areas. Water from Lake Hodges has spilled over Hodges Dam in some years but the flow down the San Dieguito River valley below the dam has been relatively mild. During the Hatfield Flood in 1916, before Hodges Dam, the water flow devastated the valley down to the ocean at Del Mar. That 1916 flood was the only "hundred-year flood" San Diego has experienced, one that is expected to occur ten times in a thousand-year period (obviously, the intervals between such floods can vary greatly).

The troublesome scenario that scares many engineers is such an extremely heavy and long-lasting rainfall coming at the end of a rainy season when the ground is already saturated, the reservoirs are close to full, and channels are clogged with debris. Heavy runoffs would then send substantial water over the spillways and downstream flooding will almost certainly occur.

Most at risk is Mission Valley, where large portions of the land are within the 100-year-flood plain, and warnings have been constantly disregarded by developers. The heavy rainfalls of 1983, for example, caused extensive flooding in the basements and first floors of many of those buildings and that rainfall was considerably short of the 100-year scenario. Also endangered are developments in the San Luis Rey, San Dieguito and Tijuana river valleys. The Tijuana River may be the best controlled of the three. Recently completed is a 3000-foot wall designed to withstand a 100-year flood. And the huge water treatment plant recently built on the Tijuana River in cooperation with Mexico is designed to withstand 500-year floods.

Frank Belock, Jr., Director of Engineering and Capital Projects for the City of San Diego, has primary responsibility for floods in the valleys below the city's major dams. Belock states that his chief concern in far South County is the Rodriquez Dam on the river, Rio de las Palmas, which flows into the Tijuana River. Delays in opening the floodgates could cause flooding problems. As to all the other rivers, neither Belock nor anyone else really knows how large the valley water flow would be today from a 100-year rainfall and what damage it could cause if it occurred at the end of a rainy season with the reservoirs already full.

Certainly a huge amount of water would sweep down all the river valleys, including particularly Mission Valley. In 1916, the Hatfield Flood totally submerged the floor of the valley, and the 1937 flood did virtually the same. Tremendous development has occurred since then and it is little comfort to reason that today's flood-conscious regulations would not permit much of that development. But Director Belock is confident that measures taken and available would control flooding, including the possible 100-year event. At worst, the City of San Diego's projections show that water in Mission Valley would rise only into basement levels of commercial buildings in the valley. Belock says there is no threat to the condominium buildings that have been erected on the San Diego River.

Environmentalists and some old-timers are not totally confident about the efficacy of existing flood control measures on the San Diego River, fearing particularly that the narrow channel funneling the river through Mission Valley will not prevent severe flooding, particularly below the Hazard Center shopping complex. One of the skeptics is O.B. "Mike" Sholders, a retired Navy captain with a master's degree in civil engineering. He has served on the SDCWA board, has been a respected consulting civil engineer in the county for many years and worked as a member of the San Diego County Flood Control Commission.

As to Mission Valley, Sholders muses, "A 100-year rainfall could very well flood parking areas for Fashion Valley, the golf course, and on west to the ocean." When asked about the San Dieguito River, he responded "Whispering Palms on Via de la Valle may be in jeopardy when vast amounts of overflow water from Lake Hodges flows down the valley, flooding residential areas west of the dam, and causing problems all the way down the valley, including flooding of Del Mar Racetrack." He also noted possible serious problems on the San Luis Rey River. "For example, that big casino never should have been built in the river." He pointed out that existing flood channels are often filled with debris kept from removal by wetland protection laws and this blockage of free flow will inevitably cause water backups and flooding.

PLANNING FOR TERRORISM AND EARTHQUAKES

America is understandably on edge about the threat of terrorism aimed at disrupting the lives of millions of people. Inside the city, Director of

Engineering Belock expresses confidence about the ability to cope with such threats but expresses particular concern about San Diego's Point Loma sewage treatment pump station that handles waste from two-thirds of the city. It is located near Harbor Drive about 300 yards from San Diego Bay, right across from the old U.S.S. *Recruit*, the stationary ship formerly used at the Naval Training Station. If this plant were blown up, raw sewage would flow directly into the bay and, Belock states, could wreak havoc for "several days" until brought under control.

Outside the city, the most obvious large-scale risk is a severance of one of the major water pipelines. This could be done by a terrorist, but fortunately all the major aqueducts are buried deep underground so their disruption is made difficult. There remains the possibility of a violent earthquake. San Diegans should note that beginning in about 1984, earthquake activity in coastal San Diego County has dramatically increased. Similar increases in earthquake activity appear to have preceded, for example, the 1906 and 1989 San Francisco earthquakes and other major shocks. Also disturbing is the fact that in the last 50 years, the rate of earthquake occurrence has doubled over that during the preceding 50 years.

In looking at the type of quake that could sever a major water pipeline, consider the Landers earthquake of 1992 that shook all of San Diego County. In intensity, it was a class IV quake with Richter magnitude of 7. There was some building damage and some water pipes were broken but fortunately not the major lines. For comparison, experts predict that it would take the far greater class VIII or IX intensity earthquakes to cause extensive building damage and break major underground pipelines.

In San Diego County, major water pipelines are in or very near four earthquake fault zones: Elsinore, San Jacinto, Rose Canyon, and San Andreas. Elsinore covers a 120-mile path across eastern San Diego from Los Angeles County to near the Mexican border. Future quake estimates in this zone are for up to 7.2 magnitude and up to class VII intensity. San Jacinto has a 150-mile path crossing the northeastern corner of the county. Forecasts are for quakes with up to 7.3 magnitude; for intensity, up to VIII in the coastal zone and up to X in the mountain areas. The San Andreas zone crosses near far northeastern San Diego County. Estimates are up to 8 in magnitude; for intensity, up to IX in East County and up to VII at the coast. For the Rose Canyon fault that runs north-south from off Oceanside down to San Diego Bay, estimates are up to 7.2 in magnitude and IX in intensity.

A book containing all of the above-cited data and more is *Earthquakes and Faults in San Diego County* (Pickle Press, San Diego 1993) written by Philip Kern, professor of geology at San Diego State University. It concludes that earthquakes of magnitude 6 and higher on the Richter Scale and up to VII or even higher in intensity are "certain to occur in the future." And in considering the possibility of critical water supply disruption, it is well to keep in mind that portions of the major pipelines carrying Colorado River and Feather River water are within fault zones not just in San Diego County but also north of the Los Angeles area; for example, the California Aqueduct carrying Feather River water runs for miles right along the San Andreas fault.

In the area north of Los Angeles, heightened fear of terrorists has caused MWD to place more guards on the ground and institute increased patrol flights around key infrastructure, including the Colorado River Aqueduct that brings water to the coast for both MWD and SDCWA. Recognizing that rapid response is critical, both MWD and SDCWA have detailed emergency plans and fully equipped emergency operations centers. San Diego planning includes full cooperation and interaction with the County's Operational Area Emergency Plan. John Economides, Chief Engineer for SDCWA, is the "Incident Manager" charged with controlling reaction to emergency situations. He said that their detailed plan is constantly reviewed and that rehearsals have demonstrated readiness to cope with major interruptions of water supplies. Asked for an overall evaluation, Economides responded, "We have made all reasonable preparations for any foreseeable emergencies."

While SDCWA is not responsible for emergency/disaster preparation by its member agencies, Economides and his staff have reviewed their plans and find them with few exceptions involving minor agencies to reflect a satisfactory level of preparation. He stresses that coordination is essential for all the county's reservoirs to be available for emergency water supply.

Noting that truckloads of material would be necessary to contaminate the water supply, SDCWA asks the public to be vigilant for any unusual trucking activity near any water facility. However, all agree that there is no sure way to keep a terrorist from tampering with the water supply system. One technological advance is a laser device to detect foreign particles in the water that was used during the last Super Bowl at Qualcomm Stadium to screen water coming into the stadium.

With respect to the critical area of communication among the many agencies including police and fire departments that would be involved in emergency situations, John Economides of SDCWA reports that great strides have been made in recent years so that radio, telephone, cell phone, and even fiber-optic links will be available to ensure ease of communication at all levels, a situation that definitely had not existed in past years.

Critical to coping with a massive disaster is the SDCWA Emergency Storage Project (ESP), designed to cope with a six-month interruption of San Diego's water supply. The prime objective of ESP, heavily financed by the current SDCWA Capital Improvement Program, is to ensure availability of enough water to maintain a 75% level of service to users in an emergency. Emergency could include a severe drought, a catastrophic interruption of imported water

. supplies, or anything that would dramatically decrease water supply to the region.

The Emergency Storage Project involves a system of reservoirs, interconnected pipelines, and pumping stations working together to store and move water when needed to all communities in the San Diego region. The Authority embarked on this $827 million project in 1989; completion is scheduled for 2010 when it will provide 90,000 acre-feet of emergency storage designed to meet county emergency needs through the year 2030. This very significant increase in reserve water supply supplements MWD's new Diamond Valley Lake near Hemet; half of that water is earmarked for emergency usage for all agencies in the MWD system, including San Diego. MWD has spent over $2 billion and labored for a decade to build this largest reservoir west of the San Andreas fault as a bulwark against earthquake or terrorist disruption of water for all of southern California.

The new Olivenhain Dam and Reservoir, first dam to be built in San Diego County for over 50 years, is a part of San Diego's emergency storage project, along with a 54-foot increase in the height of San Vicente Dam. If San Diego should be hit by earthquake or terrorist attack, the Olivenhain Reservoir's water capacity of 7.8 billion gallons is enough to sustain almost 200,000 people for a year or enough to last all three million inhabitants for two months. The new pipeline linking Olivenhain Reservoir and Lake Hodges, enabling Lake Hodges to be filled with imported water for the first time, is scheduled to be completed by 2008. Uniquely, the huge Olivenhain Dam does not block any stream or river; it simply boxes off a canyon to be filled with imported water. To permit maximum water-movement flexibility in time of emergency, Olivenhain will be linked to both the first and second San Diego Aqueducts.

Ultimately, by 2008, the completed countywide system will allow rapid movement of water supplies to meet emergencies, as when any of the major north-south pipelines should be disrupted. Water can quickly and easily be brought by SDCWA from water storage locations to usage points and moved around disrupted sections of pipeline. The flexible system will permit two-way (north-south and east-west) movement of water to handle disruptions anywhere in San Diego County.

Still to be completed is a new San Vicente Pipeline that will extend 12 miles from the San Vicente Reservoir to the Second Aqueduct that lies just west of Interstate 15. Announced by SDCWA late in 2002, the east-west route will run north of Miramar and south of Poway but has not yet been finalized. Using both open-cut trench and 12-foot-wide tunnel construction, this new 8.5-foot-diameter pipeline will enable SDCWA to respond to interruption of water delivery by quickly bringing in water from the San Vicente Reservoir. Considerable preparation work remains to be done; construction on the $200 million project is not scheduled until the year 2007.

CHAPTER 9
ALWAYS SEEKING NEW WATER

When "all those reservoirs" in San Diego were completed decades ago, most people then thought they would forever supply San Diego with its needed water. They have not. And in San Diego the water future again looked exceedingly bright when SDCWA received water from the Colorado and Feather rivers. Many thought that these rivers would satisfy all our needs. But they have not. For a variety of reasons, the search for new water is as imperative today as it has often been in the past. That search must intensify as population will almost inevitably continue to increase.

THE COMPLICATED COLORADO RIVER WATER SUPPLY

Southern California has for years depended on the Colorado River for most of its water, but very few people are familiar with the complicated formulas dividing up that water. Related earlier was the story of how MWD built the Colorado River Aqueduct to bring that water to the coast. But, as NPR's Elizabeth Arnold has recently reported, every drop of the river (and then some) has already been promised to someone, somewhere. Southern California must compete with many other users in all the Colorado River basin states for that water, which has led to a complicated series of court decisions and agreements among the various governmental agencies that rely on the water.

As noted in Chapter 4, the Colorado River Compact of 1922 divided the Colorado River between the Upper and Lower Basins, 7.5 million acre-feet (maf) per year to each. The Upper Basin states divided their share by a 1948 agreement; the Lower Basin's share was divided by a 1964 Supreme Court decree. California received 4.4 maf plus 50 percent of the surplus, Arizona received 2.8 maf plus 46 percent of the surplus, and Nevada received .300 maf plus 4 percent of the surplus. Surplus waters included both waters in excess of 7.5 maf and any portion of 7.5 maf that was unused by the Upper Basin states. In the past there has always been surplus water available for California. California is entitled to use any water not used by Arizona and Nevada to which they are entitled.

Congress gave the Secretary of the Interior the responsibility and authority to decide which users in each state would receive Colorado River water. However, it was required that water rights established pursuant to state law prior to June 25, 1929, be protected. They were known as present perfected rights and would come into play during a shortage when there was less than 7.5 maf of water available for the Lower Basin.

On August 18, 1931, as requested by the Secretary, the California entities made the Seven-Party Agreement, which recommended how to apportion and prioritize California's share. The Secretary accepted these recommendations and included them in each subsequent water delivery contract with the California entities: MWD, September 28, 1931; Imperial Irrigation District, December 1, 1932; Palo Verde, February 7, 1933; and Coachella, October 15, 1934.

Agricultural entities, which had been diverting and using Colorado River water (Palo Verde Irrigation District, the Yuma Projection California, and the Imperial Irrigation District) for more than two decades, were apportioned the first, second, and third priorities, but "not to exceed 3,850,000 acre-feet of water per annum." A fourth priority of "550,000 acre-feet per annum" was apportioned to Metropolitan for use "on the coastal plain of southern California." Thus, the first four priorities equaled the total amount of California's basic entitlement of 4.4 maf per year, and the remaining three priorities would have to be satisfied from California's right to surplus waters.

A fifth priority of 662,000 acre-feet was apportioned to Metropolitan (550,000 acre-feet) and San Diego (112,000 acre-feet). San Diego merged its water rights with MWD in 1946. A sixth priority of 300,000 acre-feet was apportioned to Imperial and Palo Verde Irrigation Districts, making a total of 5,362,000 acre-feet apportioned and prioritized. A seventh priority "of all remaining water" available for California was apportioned "for agricultural use."

PRIORITIES TO COLORADO RIVER WATER (ACRE-FEET OF WATER PER YEAR)		
Priority	User	Amount
1-3	Agricultural Irrigation Districts (3)	3,850,000
4	MWD	550,000
5	MWD	662,000
6	Imperial Irrigation District and Palo Verde Irrigation District	300,000

According to the Seven-Party Agreement, 1931

The huge disparity between the irrigation districts and the urban areas of Los Angeles and San Diego is immediately apparent. Some find it startling that Imperial Valley distributes more Colorado River water for agriculture than the cities of Los Angeles and San Diego supply to all their water users. Although such agricultural-urban water usage disparities seem hard to justify, they are dictated by the existing river law ordaining that priorities be determined by the date of filing claims, and the irrigation districts filed first.

As the years went by, Arizona and Nevada took more and more water, increasing their usage up to their priority amounts. That left only the surplus water declared available each year by the Department of the Interior and this has been relied on ever since mostly to meet the growing needs of both the Los Angeles and San Diego areas. However, this surplus has become less and less reliable. All the Upper Basin states are increasing their demands on the river so that the once plentiful "surplus" over the amounts allotted to the Lower Basin states is rapidly diminishing.

Colorado's Governor Bill Owens recently commented, "While Coloradans are carefully considering which day they are allowed to turn their home sprinklers on, California lawns are lush and green, thanks to Colorado's water." Serious consideration is being given not only in Colorado but also in Arizona and Utah to divert substantially larger amounts of Colorado River water to projects in these Upper Basin states, all of which constantly complain about California's over-usage of its Colorado River allotments.

In addition, environmental matters, such as concern for endangered species, threaten efforts to take water from the river. Farther down the river, Arizona, Nevada, and New Mexico continue to seek more and more water from the river. Mexico is also clamoring for water, seeking to take advantage of its treaty rights with the United States. On top of all this, the Colorado River is not living up to its 1922 expectations when the interstate Colorado River Compact was formed. The average river flow is now less than the total legally allotted to the seven basin states and Mexico. There is also increased concern about the quality of the water that requires more expensive treatment.

Eleven western states are deeply concerned about drought conditions that continue to plague the area. The federal Bureau of Reclamation has issued a major report entitled *Water 2025* that calls it a crisis situation and urges more intensive planning by the western states that rely on the Colorado River. Lake Powell and Lake Mead, huge reservoirs relied on for emergency water supplies, are now half empty. The *Water 2025* report concludes that the Colorado River valley running alongside both Arizona and California is especially vulnerable.

MWD has for years over-used its 550,000 acre-feet annual allotment of Colorado River water, going up to a high of 1.2 maf per year — and a considerable amount of this over-usage has gone to San Diego. Added to additional amounts used by the agricultural districts, pushing them over the 3.85 maf allowed by the Seven-Party Agreement, California's total annual usage has gone far over its 4.4 maf limit. Forced to recognize the problem as the

Department of the Interior issued repeated warnings, MWD, SDCWA, and the irrigation districts negotiated what is called the Quantification Settlement Agreement (QSA) which embodied a plan limiting the amounts that could be taken by the agricultural districts and providing for transfers among the river water users. The duration of the QSA was 15 years and the intent was to provide a "soft landing" bringing California down to its legal 4.4 million af/year limit.

Turning to the Feather River, a variety of complicated and mostly environmental factors have kept the Feather River water from living up to original expectations proclaimed by the State Water Project. A major bottleneck has been the delta where the Sacramento and San Joaquin rivers flow into northern San Francisco Bay. The delta is a triangular-shaped body covering nearly three-quarters of a million acres, consisting of a combination of salt water from San Francisco Bay and fresh water from the two rivers; some of it is 20 feet below sea level. More than two-thirds of California's drinking water, supplies for 22 million residents, passes through this delta. It also provides water to irrigate Central Valley farmland which produces an important part of the nation's fruits and vegetables.

Some of the most fertile land in California is inside the delta and over the years it has attracted thousands of farmers. Their crop production is significant and agricultural interests have vociferously opposed any major changes in water flow through the delta. There are major fisheries in the delta and the area is a sought-after haven for both fishermen and hunters. These groups are obviously resistant to interference from new water flow systems.

The delta is a major habitat for millions of waterfowl on the Pacific flyway. Over half the fish that live in the ocean and spawn in fresh water, including the Chinook salmon and steelhead trout, are dependent on delta waters. Both are threatened species under the federal Endangered Species Act. A wag has said that the Bureau of Reclamation and the State of California have found themselves "between the delta and the deep blue sea."

The "Delta Headache" has for the past quarter of a century resisted any "painkiller" solution. "It's just too damn complex," says one state legislator. The long-term solution currently being implemented is the CALFED Bay-Delta Action Plan announced by the federal and state governments in August 2000. The plan calls for both governments to cooperate in improving the quality and reliability of Bay-Delta water supplies. Included are more steps to restore the Bay-Delta ecosystem, improvement of the water conveyance system by modifying delta channels, stabilizing delta levees, fostering the transfer of water from agricultural uses, and managing the environment to improve the fisheries located in the delta. The conveyance system improvements could permit increased water flow during times of the year when additional water is available and water quality is at its highest and reduce water flow when desirable for environment protection reasons.

Legal challenges and other impediments threaten full and fast implementation of these complex plans that require steps to be taken over a seven-year period, and even that is but one stage of a 30-year program that requires action by both state and federal government. If, sooner or later, these efforts succeed in restoring greater stability to the State Water Project system, San Diego will benefit to a measurable degree. While MWD projects an additional supply of Feather River water in the future, this is far from certain and depends greatly on highly questionable implementation of the CALFED Bay-Delta program. Also, environmentalists and northern California agricultural interests remain strongly resistant to measures that would increase water flow from the delta to solve southern California's water problems. This was reflected in opposition expressed by northern Californians to a proposal advanced by MWD in August, 2003 to increase pumping of water from the delta into the California Aqueduct for use by MWD in southern California.

The original commitments by the State Water Project to its contracting parties, including MWD, have never been fulfilled. The project initially planned to deliver 4 million acre-feet to those parties which totaled about 30 agencies throughout California. MWD was entitled to almost half of the total but there is small likelihood that the initial goals can be met. SDCWA currently estimates that the chance of SWP making full deliveries is "less than 25%" and predicts that "SWP shortages are expected to become more frequent."

WILD SUGGESTIONS FOR IMPROVING
COLORADO RIVER WATER FLOW

Some have suggested (and these ideas are far from new) that southern California should look further north to far-away rivers offering vast supplies of water that can be somehow moved into the upstream stretches of the Colorado River. Theoretically, water from the Snake River could be transported from Idaho down through Nevada and into the Colorado River; $3 billion is the estimated price tag. A more expensive idea (approximately $11 billion) is to pump water from the Columbia River into Lake Mead and thence into the Colorado. Finally, soaring into the atmosphere in cost is the most elaborate proposal. Called NAWAPA (North American Water and Power Alliance), it involves tapping rivers of the Yukon and using massive systems to transport the water to the upper reaches of the Colorado River; estimated cost, $200 billion.

Critics of these extreme solutions to San Diego's water problems point out that they ignore the extremely high costs involved and the undoubted opposition of the states of Alaska, Washington, Oregon, and Idaho. The Owens Valley and Imperial Valley outcry about urban areas' outreach for water supply solutions would be multiplied many times. Finally, these critics conclude that the far-out solutions are about as practical as the old suggestion of towing Alaskan icebergs south to San Diego. So far, no one has explained how, when the iceberg gets into San Diego Bay, the melted water will be contained and economically brought ashore.

EXPANSION OF LOCAL WATER SOURCES

Until 1947, San Diego's thirst for water was sated by local water sources: rainwater runoff kept the reservoirs full, supplemented by groundwater pumped from local aquifers. But a growing population together with continued heavy military demands plus commercial and industrial expansion ended that water comfort. Imported water then filled the need as earlier described, so that by the 1980s, less than 30% (and now much less) of San Diego's water comes from local sources. But since the 1987-1991 drought, the water misers have forced the water seekers to re-focus on local sources for water. Underground water, water conservation, and water reclamation are receiving more and more attention from the water supply agencies. SDCWA has carefully tabulated the amount of water that can be generated from local sources. The outlook is for 150,100 acre-feet by 2005, 184,200 by 2010, and 198,500 by 2020. Those are healthy amounts but far from all that will be needed.

An additional source is the runoff water from rainfall that flows into the county's reservoirs, estimated at 85,600 acre-feet annually. These reservoirs constitute a portion of the county's local water supply but are primarily intended for times of water shortage and for use in an emergency. See the table, "Reservoirs in San Diego County" including San Diego on the following page.

GROUNDWATER SUPPLY

There's water under our feet, lots of it. Or so say the experts in the U.S. Geological Survey in Washington, D.C. who patiently explain that a huge underground aquifer known as the San Diego Formation is just waiting to be tapped. But they aren't really sure just where it is or whether the water in the aquifer is even drinkable. The hypothesized aquifer is composed of sand in an ancient seabed and is estimated to range in depth from 200 feet to more than 4000 feet, stretching from Mission Valley south into Mexico and roughly from East San Diego out to the ocean.

Colonel Fletcher recorded in his *Memoirs* his belief that there was an "underground reservoir under San Pasqual Valley holding 20-30 billion gallons of water." He also bragged about his wells in Fletcher Hills that tapped water down to 536 feet or 100 feet below sea level and those near El Capitan Dam that found subterranean water at an elevation of 1900 feet. His concluding sentiment on tapping underground water cannot be quarreled with: "A good well of water is more valuable in this county than oil."

RESERVOIRS IN SAN DIEGO COUNTY

Reservoir Owner	Reservoir	Capacity (Acre-Feet)
Escondido	Wohlford	6,506
San Diego	Barrett	37,947
San Diego	El Capitan	112,807
San Diego	Hodges	33,550
San Diego	Morena	50,207
San Diego	Otay	49,510
San Diego	San Vicente	90,230
San Diego	Sutherland	29,685
Helix Water District	Cuyamaca	8,195
Helix Water District	Jennings	9,790
Ramona Municipal W.D.	Ramona	12,000
Sweetwater Authority	Loveland	25,400
Sweetwater Authority	Sweetwater	28,079
Vista Irrigation District	Henshaw	51,774

Source: San Diego County Water Authority 2000 Urban Water Management Plan

Efforts are underway to find and use this underground resource for both storage and usage. Storage is the likely benefit since much of the groundwater is not suitable for drinking without expensive treatment because it contains high concentrations of salt. Studies establish a storage capacity of between 40,000 and 90,000 acre-feet, more than could be provided by an ordinary reservoir. The drawback, however, is significant. To be usable for storage, the underground aquifer would have to be first drained and this raises serious problems of massive seawater intrusion and the possibility that the land would subside. Despite these difficulties, studies conducted by SDCWA conclude that "the gross storage potential of the San Diego Formation is high" and that "there is a reasonable likelihood that cost-effective increments of regionally beneficial storage benefits can be developed."

So, the studies continue. Currently, SDCWA is investigating two of the area's groundwater basins for both storage and recovery of water. The first is the basin of the San Luis Rey River in Oceanside; the City of Oceanside also refers to this basin as Mission Basin. The second is the San Diego Formation; water is taken from it, utilizing wells at several locations from central San Diego south to the border. Funds made available by the water bond issue approved by the voters in the year 2000 could help finance this and other water recovery projects.

SDCWA member agencies are also pursuing groundwater possibilities that would offer cheaper local water and some cutback in the need for imported water. USMC Camp Pendleton and the Sweetwater Authority's National City Well Field are examples of projects extracting groundwater that does not require expensive treatment. The City of Oceanside has a substantial recovery project working with brackish groundwater. The Sweetwater Authority that supplies water to domestic and agricultural users in the National City and Chula Vista areas has the Richard A. Reynolds Groundwater Demineralization Facility. Obviously, facilities like this have considerably higher costs than those that simply extract underground water without any treatment.

Optimists are hopeful of going underground to obtain even more water. As to whether it can be substantial in amount, pessimists say the chances are slim. However, realists doggedly pursue all available alternatives. Count SDCWA among the realists; it looks at 31,000 acre-feet by 2005, 53,500 by 2010, and 59,500 by 2020.

WATER CONSERVATION

A variant on the old saying, "a penny saved is a penny earned" is the reminder that water saved is as good as new water supplied. The severe drought years of 1987-1991 and the accompanying adjurations of community leaders finally convinced the general public that it was time to act on water conservation. In her presentation *Last Oasis* (Home Vision, 1997, widely hailed by water authorities) Sandra Postel argues passionately for conservation as an absolute necessity. Noting that new sources are hard to find and delivery systems are increasingly costly, Postel concludes that a new "culture of conservation" must be accepted, recognizing that this will require "a major change in the way people think about water." Both SDCWA and its member water agencies are doing their best to get consumers on that track. Indeed, SDCWA is today recognized as managing the most aggressive water conservation program in California.

SDCWA acts both directly and through its member agencies, stating that about 187,000 acre-feet of water have been saved in the last decade and promising substantially increased savings of up to 20,000 acre-feet annually in the years ahead. On the heels of the low-flush toilets and showerheads have come ultra-low-flush toilets, pre-rinse spray heads, more watertight plumbing fixtures in homes, and "smart" weather-based water-system controllers for sprinkler systems. SDCWA's residential High-Efficiency Clothes Washer (HEW) incentive program has received the Award of Excellence from the Department of the Interior's Bureau of Reclamation. Voucher programs and free handouts are used by many SDCWA member agencies to increase usage of water-saving equipment by the public and water conservation gardening is being widely promoted. Overall, water-wise behavior by consumers is becoming more acceptable throughout San Diego County.

SDCWA is also concentrating on commercial, industrial and institutional water savings measures. Vouchers and other incentives will be offered for such devices as water-efficient commercial dishwashers, multi-load clothes washers, water-saving hospital X-ray film processors, and water-pressurized brooms. Many outdoor opportunities are stressed, including education and training programs aimed at more efficient water usage, replacement of grass sports fields with artificial turf, installing water re-circulating systems, free irrigation system evaluators for agricultural properties and micro-irrigation programs for grove watering.

Considerable progress was made during the 1990s when publicity and peer pressure achieved conservation results in many San Diego communities. Water usage is finally no longer increasing at the same rate as the population, but many more conservation efforts by the water-consuming public are still needed. The following table shows the SDCWA projections of amounts that need to be conserved as shown in its 2000 Urban Water Management Plan. Certainly they are optimistic, but remain achievable goals given widespread public participation.

PROJECTED WATER CONSERVATION SAVINGS
(ACRE-FEET OF WATER PER YEAR)

Year	Amount
2005	54,900
2010	74,400
2015	83,400
2020	93,200

Source: SDCWA 2000 Urban Water Management Plan

THE PURPLE PIPE PROJECTS: LOCAL WATER RECLAMATION

Every gallon of recycled water used means one less gallon of "new" water that San Diego needs to import. SDCWA knows this lesson well and is actively encouraging all its member agencies to intensify water reclamation projects utilizing recycled water, meaning water treated and disinfected to be suitable for non-potable water consumption. SDCWA has developed a new piping system in which recycled water will be carried to industrial parks and recreational areas where the water can be put to good usage, thereby freeing up more pure water for domestic uses. This is SDCWA's "purple pipe" system.

The Santee treatment plant is a notable example of public-use water reclamation. This project, created by the Padre Dam Municipal Water District, used wastewater recycled to create Santee Lakes as a part of Santee Lakes Regional Park, a recreational area for people to enjoy. Golf courses are another target use for recycled water. Huge sprinklers throw many thousands of gallons of precious water onto the golf courses so golfers can have a better "lie" in the fairways and a smoother putt on the greens.

Other examples of recycled water use include landscape irrigation, filling lakes, ponds and ornamental fountains, watering campgrounds, freeway medians, community greenbelts, school athletic fields, nursery stock, and controlling dust at construction sites. At a newly constructed detention facility at Otay Mesa and a biotechnology firm in La Jolla, dual-plumbed facilities have been installed to allow use of recycled water for toilet and urinal flushing. Along with other water districts, the Otay Water District has historically used reclaimed wastewater for irrigation and other non-potable uses including landscaping and watering of golf courses and playing fields used by schools in the district. It recently concluded a deal with the City of San Diego hailed by Mayor Murphy as "the biggest recycled water deal in San Diego County history."

Otay will purchase millions of gallons of non-drinkable recycled water from the city's South Bay Water Reclamation Plant for irrigation, commercial, and industrial usage. This $150 million plant is in the Tijuana River valley right next to the International Wastewater Treatment Plant just north of the border. The water will sell for $350 per acre-foot, about half the cost of potable tap water. Potential customers include the Otay Water District, Caltrans, and the Border Patrol.

The City of San Diego's North City Water Reclamation Plant near Interstate 805 and Miramar Road in North County, overlooking University City and called "the jewel of the city's reclamation program," handles 25 million gallons a day of sewage, reclaiming about 4 million. The virtually clear and odorless water, fit for swimming but definitely not drinking, is sold to about 100 customers, including the municipal Torrey Pines Golf Course and General Atomics. Another City plant is in South County, the South Bay Water Reclamation Plant.

SDCWA, working with its member agencies, has developed a major reclamation program that provides up to 25% of the costs of wastewater recycling, and includes numerous specific projects. The program partners that include San Diego City, Escondido, Poway, Otay, Padre Dam, Sweetwater, and Tijuana Valley have developed a system of interconnected water recycling projects.

There are, regrettably, built-in limits on the use of recycled water. Conventional treatment processes are not designed to screen out any dissolved substances. Therefore, only suspended particles are removed. As salinity content

Santee Lakes Regional Park   *Padre Dam Municipal Water District*

increases, usability of water for agricultural purposes decreases. So, for example, recycled water cannot be used for some crops and for nursery stock. Additional treatment of the water is required. Finally, most reclaimed water cannot be conveyed through the same pipelines as drinking water and the "purple pipes" that carry it cost $1 million or more per mile to lay. This severely limits the ability to expand usage of recycled water.

Overall, public education and hard work by involved agencies promise significant water supply gains from expanded usage of recycled water. Recycled usage accounted for about 13,700 acre-feet in 2000. SDCWA projects annual increases of 33,400 acre-feet by 2005, 45,100 by 2010, and 53,400 by 2020.

The reverse osmosis process for removing salt from water has the virtue of substantially low energy costs since the water does not have to be heated to high temperatures for evaporation. Molecules of salt and other minerals are removed by pushing the salt water under high pressure through a filtering membrane. Small-scale reverse osmosis plants have existed for years in Mission Valley just west of Qualcomm Stadium.

Oceanside operates the Mission Basin Desalting Facility that takes brackish water from Mission Basin, a groundwater basin also known as the San Luis Rey River basin, and recycles it for domestic usage in 4000 households. Oceanside is currently considering a major expansion of the facility, reaching for production of 5 million gallons of usable water per day.

THE OCEAN IS AT HAND — WHY NOT DESALINATION?

The Pacific Ocean is literally at our doorstep, so why not a massive desalination plant? Traditionally and for many centuries, saltwater has been converted into drinking water by a thermal process, simply heating the water until evaporation takes place. The evaporated water (steam), clear of all non-volatile impurities, is then condensed into pure water. Distillation, it is called. Ships at sea and large plants around the world have for many years used distillation to obtain drinking water. The Middle East has many such facilities and the largest distillation facility in Saudi Arabia treats 143,360 acre-feet per year. Around the world, desalination facilities are giving the lie to Coleridge's Ancient Mariner, who chanted, "Water, water everywhere, Nor any drop to drink."

Environmentalists are raising some questions about desalination in California. The California Coastal Commission staff has recently warned about possible harm to the Pacific Ocean environment. But the major drawback has consistently been cost. As is so often the case, the issue is economics and desalination has historically been a budget-breaker. Not just the costly equipment, plants, and operating manpower but also the increasingly expensive energy needed to raise water temperature to the evaporation level. And, by no means least, the costs of pumping the water to the point of usage and disposing of the waste that results from the operation. To use one comparison, it takes about 5 barrels of oil to pump 1 acre-foot of State Project water into southern California and about 3 barrels of oil to import the same amount of water from the Colorado River. Fifty barrels of oil are required to produce the same amount of water from desalination through distillation. And production is not the end of the costs; the water must still be moved from sea level to distribution facilities.

Michael McClary once wrote, "Irrigation of the land with seawater desalinated by fusion power is ancient. It is called rain." Seriously, though, costs have forced consideration of cheaper atomic power for desalination. A nuclear power plant would substantially cut costs. So, San Diego has looked at the San Onofre Nuclear Generating Station near San Clemente and MWD has seriously considered a proposed Bolsa Island project that would alternatively involve fossil-fired or nuclear-powered facilities. These alternatives sound attractive but there are difficult bridges to cross before they could become a reality. One is the public concern about nuclear power plants; whether justified or not, it must be taken into account. And another fully justifiable concern is safe disposal of the spent fuel.

In the early 1960s, one of the first desalination plants in the nation was built by the Department of the Interior on land supplied by the Navy on Point Loma near the Cabrillo Monument. Using the distillation process, the plant produced 1 million gallons per day at a cost of $1 per thousand gallons and the water was sold to the city for 20 cents per thousand gallons. The experiment ended in 1964 when Fidel Castro cut off the water supply to the Navy installations at Guantanamo Bay, Cuba, and the plant was dismantled and sent to Cuba.

The high cost of distillation plants has put the desalination spotlight on reverse osmosis; new plants using this technique are becoming common. In

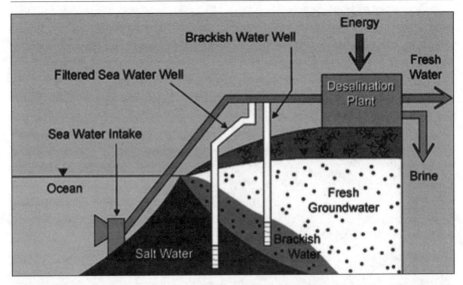

Desalination schematic   *City of San Diego Water Department*

Florida, Tampa Bay's new $110-million plant is the largest in the Western Hemisphere, supplying 25 million gallons of fresh water per day and its experience indicates that the cost of seawater desalination may be decreasing. In Hawaii, several Japanese companies are engaged in research and development and one company, Koyo USA Corp., expects to have a plant operational in the near future. In California, smaller plants are in operation at Catalina Island and Hearst Castle. Trinidad has quoted a price of $865 per acre-foot at its plant but San Diego's imported water costs only a fraction of that amount. In 1967, a pilot plant was set up in South Bay using power from San Diego Gas & Electric. It was capable of doubling the performance of the old Point Loma plant. However, the costs were still prohibitive.

In November 2002, SDCWA announced a partnership with Poseidon Resources to build a $270 million desalination facility on the coast near Carlsbad, next to the Encina Power Station. Its 56,000 acre-feet annual capacity would make it the largest desalination facility in the Western Hemisphere. In its Regional Water Facilities Master Plan, SDCWA included substantial investments in this facility and in May, 2003, approved a $1.5 million contract for a feasibility study on building the system to convey the water into San Diego's existing pipeline system. For the Encina plant, SDCWA projects a supply of 50,000 acre-feet by the year 2010, increasing to 80,000 by 2015. The Authority is also evaluating successor sites; obvious choices would be the South Bay power plant at Chula Vista which already has seawater intake facilities and the San Onofre Nuclear Generating Station near San Clemente.

Recognizing the need for additional funding, SDCWA has gone to Washington for help. San Diego area congressmen Randy "Duke" Cunningham and

Encina Power Station location for desalination plant
San Diego County Water Authority

Duncan Hunter have introduced legislation that would appropriate $750,000 for the SDCWA seawater desalination program. This would supplement other proposed legislation sponsored by the U.S. Desalination Coalition establishing a nationwide competitive grant program.

MWD and its member agencies to the north have invested considerable sums in desalination plants and plan to invest even more. There are the Orange County plant, the Los Angeles Scattergood plant, a Long Beach facility, and one sponsored by the West Basin Municipal Water District plant. Huntington Beach is currently considering a new desalination plant. President/CEO Ron Gastelum of MWD stated in a recent interview that both board members and staff personnel have worked closely and cooperatively with San Diego on plans to intensify usage of desalinated ocean water. Encouragingly, a 27-member state task force in California announced in late 2003 its findings that desalination has indeed become more economically feasible and advocated increased state funding for further research.

WHAT IS THE BOTTOM LINE?

As San Diego continues to grow, water usage spirals upward. Hopefully, demand will be lessened as the conservation practices previously discussed continue to take effect. The following table shows that demand with conservation estimates included.

ESTIMATED WATER DEMAND
(ACRE-FEET OF WATER PER YEAR)

Year	Amount
2001	674,000
2005	697,326
2010	722,104
2015	760,795
2020	801,743

Source: SDCWA 2000 Urban Water Management Plan and SDCWA Regional Water Facilities Master Plan, December 2002

Thus, San Diego County's water demand is expected to be about 800,000 acre-feet in 2020 when the population is expected to be about 3.85 million, a substantial increase over today's roughly 3 million.

The following table shows the projected increases in local water supplies obtained from runoff (surface water), groundwater, recycling, and desalination.

PROJECTED LOCAL WATER SUPPLIES
(ACRE-FEET OF WATER PER YEAR)

Year	2005	2010	2015	2020
Surface Water	85,600	85,600	85,600	85,600
Recycling	33,400	45,100	51,800	53,400
Groundwater	31,100	53,500	57,500	59,500
Desalination	0	50,000	80,000	80,000
TOTAL	150,100	234,200	274,900	278,500

Source: SDCWA 2000 Urban Water Management Plan and SDCWA Regional Water Facilities Master Plan, December 2002

Assuming a demand ranging up to about 800,000 acre-feet annually by the year 2020, the deficit between local supply and demand is obviously very substantial and could become even greater if a series of drought years were to occur, decreasing substantially the amount of local water available to meet demand.

To fill the gap, San Diego must continue to rely heavily on imported water supplied by MWD from the Colorado River and California Aqueducts. SDCWA has projected the amounts it can conservatively count on receiving from MWD in the December 2002 draft of its Regional Water Facilities Master Plan. These amounts, labeled San Diego's "preferential MWD water rights," are listed in the following table.

PREFERENTIAL MWD WATER RIGHTS

(ACRE-FEET OF WATER PER YEAR)

Year	Amount
2005	344,800
2010	368,100
2015	345,895
2020	313,243

Source: SDCWA 2000 Urban Water Management Plan and SDCWA Regional Water Facilities Master Plan, December 2002

SDCWA consistently obtains much more imported water from MWD than these amounts but conservatively relies on these figures in computing supply versus demand, since technically MWD has the right to reduce SDCWA's "take" of imported water to these levels at any time its board so decides. As previously discussed, that was the power given to MWD by the Section 135 formula for computing preferential rights to water by its member agencies.

A glance at the local water supplies and expected demand figures given earlier shows quickly that San Diego definitely cannot count on meeting all its needs with its existing preferential rights to water from MWD. The deficit between supply and demand is substantial, ranging up to about 200,000 acre-feet per year. To date, this difference has been met by MWD supplying imported water far above San Diego's preferential rights. Currently, MWD supplies over 600,000 acre-feet annually, about 90% of total demand.

This situation has understandably disturbed the SDCWA board of directors who have embarked on ambitious programs to substantially decrease San Diego's reliance on imported water purchased from MWD. To drive the percentage from 90% down to a considerably more acceptable 40% would require much more water than can be supplied by increases from surface water runoff, conservation, recycled water, and groundwater aquifers. New programs have become absolutely essential and the one that has offered the most promise is acquiring water now wasted in the Imperial Valley agricultural irrigation programs. "Water transfer," it is called.

CHAPTER 10
THE AMBITIOUS AG-URBAN
WATER TRANSFERS

For years, city folks driving for hour after boring hour along Interstate 5 through the agricultural heartland of California have seen huge sprinkler systems and miles of river-like canals and irrigation ditches bringing water to thousands of acres of fields and orchards that today produce a very large portion of the nation's staple foods. Many of those drivers have wondered about the state's water system and asked, isn't there an imbalance between agricultural use and urban use? Why are the cities in southern California clamoring so loudly for water when it seems that so much is being wasted in moving it to those plants and trees? What happens to the water that remains in the irrigation ditch when it reaches the end of a row of plants? Aren't there a lot of leaks? In the open ditches and canals, how much seeps into the ground?

These questions became much more pointed when southern California heard the "wake-up call" generated by the 1987-1991 drought. And to answer the questions, water experts have increasingly considered the admitted waste in agricultural irrigation systems. The California State Water Resources Control Board found in 1984 that irrigation districts have been wasting water and using it unreasonably in violation of California constitutional prohibitions. The federal Department of the Interior questions whether the old "reasonable and beneficial use" water doctrine is being violated. Some have said that up to half of the water going into irrigation systems is lost, never used for watering plants and trees. Why not capture that lost water and transport it to the urban centers of the state?

All who carefully consider the situation agree that enormous water savings can be painlessly realized by implementing end-user water conservation measures in agricultural areas and that huge amounts of water thereby "saved" can be moved to water-needy urban areas. Now accounting for only a small part of urban water usage, water transfers from agricultural waste to urban usage — "ag-urban transfers" — will grow exponentially in coming years.

The process involved sounds simple but can become enormously complicated. The urban user pays to the agricultural district an agreed-upon amount per acre-foot of water conserved and transferred, and that money is used to compensate farmers for their conservation practices. Since the water must be transported to the point of usage through one of the major aqueducts such as the California Aqueduct or the Colorado River Aqueduct, the user must pay for the cost of transportation, an amount that is termed a "wheeling charge." California law requires that water agencies controlling the aqueducts must allow the wheeling to take place in exchange for fair compensation for the use of the system. There is, however, considerable debate as to what "fair" means and the total amount paid must be taken into account by the urban user in assessing the economics of utilizing this transferred water.

IMPERIAL VALLEY POTENTIAL FOR AG-URBAN TRANSFER

San Diego's efforts to diversify its water supply sources long ago led SDCWA to seriously evaluate agricultural-urban water transfer possibilities. For years, San Diego had looked with covetous eyes on Imperial Valley water and finally approached valley officials in 1988. In that agricultural paradise, 570,000 acres of farmland with mild winter temperatures have four-season crop production. The valley is a major food supplier for the nation; about one-seventh of America's winter vegetable crop is supplied from Imperial Valley. The valley uses up to 70% of California's 4.4 million acre-feet allotment of water from the Colorado River. And drainage of that water from the many thousands of acres of farms flows unused into the Salton Sea.

Substantial funds will be needed to complete the concrete lining of the All-American Canal and the Coachella Canal in Imperial Valley; recently proposed additions would bring the total to over $200 million. As this program is put into effect, huge amounts of water lost through seepage and leaks will be conserved. This solution, however, has not made all farmers happy; those south of the border complained that their wells would dry up as the underground water flows from canal seepage disappeared.

In 2002, the Imperial Irrigation District (IID) diverted and distributed about 3.2 million acre-feet of water from the Colorado River. Its irrigation system includes almost 1700 miles of delivery canals, numerous reservoirs, over 1400 miles of drain ditches, and almost 34,000 miles of tile drains (drain pipes beneath the irrigated fields that allow saline water to be carried away from the root zone). It has long been recognized that this system provided opportunities for the recovery of wastewater not delivered to crops.

MWD was the first to capitalize on this opportunity for "new" water, negotiating in 1988 an agreement with IID to transfer in future years 100,000 acre-feet of water. In 2002, MWD agreed with Palo Verde Irrigation District to take up to 111,000 acre-feet of agricultural wastewater each year over a 35-year period.

But by far the biggest water transfer was initiated by SDCWA. It optimistically aimed at 200,000 acre-feet annually, far greater than the 112,000 acre-feet that was San Diego's original share of Colorado River water, making a huge dent in the annual acre-feet over-usage by California of its share of Colorado River water and providing enough to supply about one-third of San Diego's annual water needs.

SDCWA now buys all of its imported water from MWD and this supplies up to 90% of San Diego County's needs. By 2020 if the water transfer plan is fully effectuated, the percentage will drop substantially. No wonder that Bernie Rhinerson, Chairman of the SDCWA board, calls the water transfer agreement with IID "the centerpiece in the Water Authority's program to diversify and improve the reliability of San Diego County's water supply."

THE SAN DIEGO-IID WATER TRANSFER AGREEMENT

Under the agreement as initially conceived, IID and its agricultural customers would conserve water and sell it to SDCWA for at least 45 years. Unfortunately, many people in Imperial Valley vehemently opposed the agreement after it was first announced in 1988. Wally Leimgruber, Imperial County Supervisor, predicted in mid-2003 that 90% of the valley residents would vote against the transfer plan if it were placed on the ballot. Tempers have flared at public meetings. Bumper stickers proclaimed, "Not another Owens Valley Rape." The agreement was repeatedly savaged as constituting a "kneeling before the politically powerful, sacrificing an agricultural region's birthright that is being taken away by greed and power." "Why," opponents screamed, "is our water going over to San Diego's golf courses?"

For a variety of reasons, some plain and some obscure, the powerful MWD from the outset took positions on the San Diego water transfer agreement ranging all the way from "lip-service support" to a "hands-off" stance to undercover opposition and even to outright opposition. Dennis Cushman, Assistant General Manager at SDCWA and frequent "point man" for the Authority on the agreement, has been in the past openly critical of MWD: "Publicly, they sometimes support San Diego's water transfer agreement, but really they do not." Observers in Imperial Valley agree and often go even further. Many have said unequivocally during the long dispute over the water transfer that in their opinion MWD not only did not want it to succeed but hoped that failure would open the door to a new MWD deal to obtain the water.

MWD's at the least ambivalent and sometimes openly hostile attitude toward the San Diego agreement was hard to square with its 1991 *Water Odyssey: The Story of Metropolitan Water District,* wherein its own very similar Imperial Valley water transfer agreement was hailed as "an enlightened approach, indeed a model, for making more effective use of limited water supplies."

And MWD had publicly taken credit for numerous transfer agreements that it had negotiated with other agricultural water districts; in the Sacramento Valley alone, MWD pursued such agreements with 14 water districts.

The San Diego agreement could have been viewed by detractors from the beginning as threatening MWD's role as imported water supplier for southern California. "Face it," said James Taylor, assistant general counsel for SDCWA who has worked since 1995 on the San Diego agreement, "Metropolitan has always had an institutional mindset that it should be the sole purveyor of imported water for our region."

Soon after the San Diego agreement possibility was announced, MWD waged a quiet but intensive battle in Sacramento against San Diego's plan. According to Dennis Cushman of SDCWA, no less than $2 million was spent by MWD on lobbying efforts to undermine San Diego's position, and MWD member agencies supporting the MWD position paid thousands of dollars to private investigators to get information that would discredit SDCWA directors and others supporting the water transfer deal. Showing up in the dossiers compiled by the investigators were alleged links with the unpopular Texas Bass brothers who had bought substantial land in Imperial Valley and were disliked as a symbol for outsiders acting against what local farmers thought were the valley's best interests.

Cushman has pulled no punches in summarizing the MWD activities in Sacramento: "MWD clearly used vicious and unscrupulous tactics in trying to defeat the San Diego-IID water transfer." Outraged legislators accused MWD of being "out of control," introducing legislation that would alter MWD's power structure. MWD backed off but some knowledgeable capitol observers refused at the time to believe that MWD's opposition to the San Diego plan had vanished.

ROOTS FOR IMPERIAL VALLEY HOSTILITY

The roots for the hostility by many Imperial Valley farmers to the water transfer agreement run deep. Giving up water, any kind of water, is anathema to many of them who view that Colorado River water as their birthright. Said Michael Morgan whose family farms 7000 acres, "When my grandfather gave his water rights to the district, he never thought he'd lose control of them." Also, there is an historical bias by old-timers against San Diego: "Those folks over in San Diego by the sea just want to use us whenever they can. When the weather is nice, the fishing in the Salton Sea is plentiful, and the dove hunting by the irrigation canals is good, here they come. When times get tough, there they go back over the mountains and ignore our problems." There is also a longtime resentment at a perceived refusal to give Imperial Valley due credit for its role in bringing about the Colorado River developments that have saved the west from water supply disaster.

In the 1990s, with agricultural production in Imperial Valley at its peak, the Bass brothers out of Texas — Sid, Edward, Robert, and Lee — bought 43,000 acres of land in the valley, stating that they wanted to run cattle. But it soon became apparent that they were more interested in selling water than cattle, stating publicly that they supported proposals to transfer conserved agricultural water to Los Angeles and San Diego. True, SDCWA did flirt with the Bass Brothers. Steve Erie, the UCSD professor previously mentioned, who has written extensively on water issues and has testified as an expert witness on behalf of MWD, has described "secret meetings" that officials of SDCWA had with the controversial Bass brothers in an effort to negotiate purchases of Imperial Valley water.

Adverse publicity about those "secret meetings" helped persuade the Bass brothers to leave Imperial Valley. "From that day," said Larry Bratton, an El Centro shopkeeper, "fear set in." For whatever reason, the Bass brothers sold their interests to U.S. Filter, part of the huge Viendi water conglomerate that includes Culligan Water, and U.S. Filter was not involved in the water transfer dispute. However, the Bass brothers memory still rankles.

Add to this volatile mix the fact that the Imperial Valley population has a very large percentage of men and women who have worked in the fields, people who are viscerally hostile to anything that threatens agricultural jobs in the valley. And fallowing or idling agricultural land, taking it out of production to save irrigation water, has consistently been a part of the water-saving process required by the transfer. That means, say the most vocal of those opposed to the agreement, a loss of jobs. The final factor in valley hostility to the water transfer agreement has always been the Salton Sea. Some in the valley predicted dramatically that the agreement would kill the Salton Sea and the Salton Sea would kill the agreement.

PIVOTAL ROLE OF THE SALTON SEA

In Imperial Valley, memories are strong of the origins of the Salton Sea. Previously described was the rampaging Colorado River's breakthrough that inundated the Colorado Desert in 1905, flooding large portions of the valley and creating the sea in the deeply depressed bed, below sea level, of ancient Lake Cahuilla. Today, covering 381 square miles, it is California's largest lake. At the same time that Imperial Valley was growing to become a "market basket" for the nation, Salton Sea became California's most productive fishery as well as a highly touted vacation destination. Travel writers waxed poetic about great blue herons skimming over breeze-blown waves while western sandpipers foraged near the shore. In the 1980s, spots like Johnson's Landing attracted thousands of eager fishermen and a 32-pound, 40-inch corvina is still mounted on the Landing wall. Fishermen bragged about pulling in tilapia fish as fast as they could throw out baited hooks from boats and docks.

Bombay Beach, on the northern shore of the Salton Sea, was written up as a "dream city in the making" as real estate agents prospered. Salton City expanded rapidly as a residential haven as subdivisions were plotted and lots were sold by the thousands. Inlets were carved in the shoreline and homeowners built small docks for their fishing boats. Famous visitors came from Hollywood and the East Coast. Movie stars and rock idols played on championship golf courses, caught trophy fish, and watched world-class speedboat races, entranced by the fact that the heavily salted water made possible the fastest boating to be found anywhere in California.

Then, as salt and nutrient concentrations increased in the waters of the sea, it "turned sour." The nutrients are "fast food" for the fish and they thrived. However, the summer of 1998 brought huge bursts of algal (from algae) blossoms fed by the growing nutrient concentrations. Then, thousands of the bursting blooms died and rotted. The decomposing blooms absorbed so much oxygen from the water that there was no longer enough for the fish. Death then worked its way up the food chain to the birds. As millions of tilapia and croakers died, so eventually did beautiful brown pelicans and snowy plovers.

As with the giant oil spills in the ocean, pictures of masses of rotting fish and dead birds being washed ashore filled TV screens and newspaper front pages. A 1999 story read, "Welcome to the Salton Sea, where the fish are rotting, the birds are dying, the algae is blooming. An All-American story about an ecosystem gone haywire." The science magazine *Discover* called the sea a "fluid contradiction of promise and poison, a death-trap teeming with life." Salton Sea was labeled a "wildlife morgue." Newspapers dredged up old stories about the Salton Sea having been used as a target for dummy bombs dropped by B-29s like the famous *Enola Gay* in preparation for the devastating Hiroshima nuclear attack in World War II. Not a plus, definitely, in the losing public relations battle being fought by the Imperial Valley Chamber of Commerce.

This has been the age of the environmentalists and they promptly took public notice of the happenings in the Salton Sea, quickly seizing on the plight of some species of birds and fish that were close to being labeled as "endangered species." The Audubon Society made Salton Sea a high priority; Salton Sea became a lightning rod for the environmentalists. Repeated constantly in environmental publications was the fact that the California wetlands used by the water birds on the Pacific flyway were being attacked by the machines of land developers, and as a result more than 90% of the coastal wetlands were disappearing. These provide the stopping or resting points for millions of migrating birds on the flyway and environmentalists warned that the Salton Sea, which had become the most important interior water body on the flyway, might be lost as a refuge.

The Salton Sea has the second highest count of different bird species in the nation, well over 400. And for some, the sea is critical. Salton Sea has 95% of the North American population of eared grebes, 50% of ruddy ducks, 80% of white pelicans, and 40% of Yuma clapper rails. Other sensitive species include brown pelicans, white-faced ibis, snowy plovers, and razorback suckers. Some bird enthusiasts call the Salton Sea "California's crown jewel of avian biodiversity." "Preserve the Salton Sea!" became a battle cry for the environmentalists as well as many people of Imperial Valley who were watching the destruction of a local dream.

The Salton Sea Reclamation Act was passed by Congress in 1998, thanks to a Congressional appropriation largely engendered by Congresswoman Mary Bono (widow of Congressman Sonny Bono, the popular entertainer who became the mayor of Palm Springs before going to Congress). However, it authorized only a fraction of the funds needed to accomplish the proclaimed goal. The act did direct the Secretary of the Interior through the Bureau of Reclamation to report to Congress with a feasibility study on Salton Sea restoration by January 1, 2000. The study has never been provided. California Proposition 204 in 1996 did provide state funding to match federal funds for such projects.

The Salton Sea Authority, created in 1993 by the Coachella Valley Water District, the Imperial Irrigation District, Riverside County, and Imperial County, has for years worked on a number of pilot restoration projects. Various desalination measures continue to be evaluated, one involving use of the geothermal plants that surround the Salton Sea to provide the energy consumed in the desalination process. One major corporation has proposed creating a huge "donut" in the middle of the lake, pumping salt water into the "donut hole" where evaporation will leave fresh water that can be moved out to the periphery of the circle.

People in the valley have for years held onto hopes for Salton Sea restoration so that it will again become an economic engine bringing tourists, fishermen, outdoor enthusiasts of all kinds, businesses, residents, and money to the area. People throughout the valley who will benefit from the restoration are, as some put it, "seeing light at the end of the tunnel." As Norman Niver, a long-time resident and member of the Salton Sea Authority stated, "If we have the water, if we have the sea, we have hope." While many in the valley saw the San Diego water transfer agreement as a threat to fulfillment of those hopes, others hailed it as a way to help pry substantial funds out of Sacramento to help solve the Salton Sea's woes.

THE WATER TRANSFER AND THE FATE OF THE SALTON SEA

Intertwined is perhaps the best word to describe the relationship between fulfillment of the San Diego water transfer agreement and the fate of the Salton Sea. The agreement was predicated on conserving water used for agricultural purposes and then transferring the water saved to thirsty San Diego. First, farmers are persuaded to adopt water conservation measures. Examples: "pump back" or collecting unused water at the end of the irrigation row and pumping it back to the beginning where it can again flow down the watering ditches; lining ditches with concrete; "dead-level irrigation," with such a slight gravity flow that practically no water remains after slowly traveling down the ditch next to the plants; installation of check-gates to eliminate leakage; utilization of lateral interceptors by farmers to create pond water that can be saved for use only when needed.

The amount of water to be conserved by individual farmers is not quantified in the water transfer agreement, nor is the amount actually saved computed and reported at any point. Also, the amount of water that could be saved from seepage as a result of lining the All-American Canal is independent of this agreement. The heart of the agreement can be simply stated. IID agrees that the total specified amount of Colorado River water in acre-feet annually will be diverted so that it does not flow into the All-American Canal for use by IID.

SDCWA will pay for that water an agreed amount per acre-foot — about $250, a large sum when compared with the $16 per acre-foot that the farmers pay for their water. Water equal in amount and quality is to be diverted from the Colorado River upstream into the Colorado River Aqueduct owned and operated by MWD. Then, there will be a "wheeling" charge by MWD for transporting the water through its Colorado River Aqueduct and delivering the same amount of water to SDCWA's aqueduct in San Diego County. The net effect is that IID takes less water from the Colorado River and the difference goes to SDCWA for which it pays IID. IID is responsible for reimbursing the farmers for their conservation practices.

CLAIMED ADVERSE EFFECTS OF
THE TRANSFER AGREEMENT

Salton Sea is termed a "terminal lake," meaning that it has no outlets. It is, in effect, a wastewater sump that has virtually no other replenishment source of real consequence. Imperial Valley agricultural water runoff keeps the Salton Sea from evaporating. Today, the sea is close to outgo-income balance, meaning that water evaporated is roughly equaled by incoming wastewater from irrigation. Opponents of the water transfer plan argue that the balance will become imbalance as water conservation measures cut the amount of wastewater flowing into the sea. Thus, it is claimed, the water level will inevitably fall as the agreement is implemented.

This triggers the so-called "domino effect." A drop in the water level results in more concentration of dissolved salt in the water. Some experts say that the salinity level has recently been brought under control and has for some time hovered steadily at a high (now exceeding that of ocean water) but still non-dangerous level. Others, however, contend that the salinity is still increasing and could threaten fish life within 20 years regardless of the effects of the water transfer agreement. All agree that increased salinity can at some point threaten fish reproduction.

When salinity kills the fish, bird life is threatened because their food (fish) is not there. Already a problem at the Salton Sea resulting from diseases, bird die-offs could be vastly increased. Also, since the sea is very shallow, shorelines are exposed as the water level goes down and deposited salt remains on the land. Computations show that if inflows of water were reduced by 200,000 acre-feet, as much as 75 square miles of lakebed could be exposed. Obviously, this has an adverse effect on those with homes on the sea.

More serious is the fact that strong winds sweeping over the naked, salt-encrusted land may kick up minor salt and nutrient particles, creating a dust permeating the air that is trapped by the prevailing northwesterly winds against the Chocolate Mountains hugging the east side of the Salton Sea. Air quality in both Imperial Valley and Coachella Valley, already marginal, could be made worse. Opponents of the water transfer agreement do not hesitate to conjure up images of the "dust bowl" that some say Owens Valley became after that valley's water was transferred down to the City of Los Angeles.

The drying up of the Owens Lake has indeed created dust storms so large that the region is considered one of the dustiest place in the United States by the U.S. Environmental Protection Agency. Opponents point out that reduction of the Salton Sea as a result of water transfers could expose two to three times the acreage involved in Owens Valley. Suffering most from dust storms would be people with asthma and chronic bronchitis, precisely the disabilities that afflict many elderly people living in Imperial Valley and in the numerous Coachella Valley communities north of Salton Sea.

Even though these ill effects cannot happen immediately and a period of years would be required for any such environmental disaster to build up, people dependent on the Salton Sea are upset about the problems. Norman Niver has a home on the sea and argues that "we are the forgotten orphans." He describes the resentment felt by those who have invested their life savings in Salton Sea properties and have seen numerous promises broken. A drive around the shores or a boat trip around the sea reveals dramatically both wasted recreational opportunities and run-down facilities.

The environmental scenario was summarized on January 18, 2003 in an Op-Ed piece in the *New York Times* by Antonio Rossman who teaches land use and water law at the University of California at Berkeley. Rossman stated with obvious exaggeration that "every acre foot of water transferred to San Diego would shrink the Salton Sea" and that the transfer would have "terrible consequences." A footnote about the author reveals that he was "special counsel to Imperial Valley," meaning the Imperial County Board of Supervisors that for years opposed the water transfer agreement, so his negative views were perhaps understandable.

There is also the job impact from any land fallowing measures that could flow from the agreement. Idling of land means loss of agricultural jobs. "'Fallowing' is a dirty word to many farm workers in Imperial Valley," stated Dave Nuffer, a local historian. It is sometimes even referred to as the "unspoken 'f ' word." Opponents of the water transfer plan have asked why jobs should be sacrificed to send water to San Diego to be used for golf courses and artificial lakes. Also frequently raised by opponents was the fear that IID and even individual farmers could be sued if the adverse consequences described above should materialize; for example, by those exposed to health hazards. In the negotiations, IID representatives repeatedly asked for indemnification provisions, an agreement that the government would pay for any such damages incurred by IID and Valley farmers.

ARGUMENTS SUPPORTING THE WATER TRANSFER

Those who have supported the agreement protest that the "terrible consequences" pictured in the above-described scenario were greatly and falsely exaggerated. First, they say "Whoa, not so fast!" The amount of water conserved in the earlier years of the water transfer agreement is small; 10,000 acre-feet in the first year and the ramping-up is gradual thereafter so that it will take ten years to reach the ultimate amount of 200,000 annually. Thus, it would take years for there to be any substantial effect on the Salton Sea from the slow-down of incoming agricultural water. Much of that comes from the huge layer of drainage tiles that lay under the land, 8 to 10 feet down. No overnight, no near-term curtailment of water flow into the lake can occur from a slowing in the amount of drainage coming through that system and into the Salton Sea. It was also pointed out that the sea can handle reductions in water level up to three feet without adverse consequences.

In addition, proponents of the water transfer agreement have pointed out that some of the salt in the sea regularly precipitates or falls to the bottom. This slows salinity increases in the water and postpones any toxic effect from water level decreases. Finally, nature itself may solve the problem. For example, when Hurricane Kathleen swept through Imperial Valley in 1976, this tropical storm substantially increased the water level at Salton Sea. Even a few years of heavier-than-usual rainfall would materially raise the water level.

Regarding the fallowing of land as initially proposed, proponents argued that it would be limited in amount and that there would be far less water flow control from this than the other proposed water conservation measures. No one argued that there could not be some job losses to those who do the stooping, picking, and carrying for minimum wage pay. However, they did say that the job losses would be minimal and involve almost entirely people who routinely travel the crop route through California so that it is nearly impossible to tell how much the idled land due to fallowing would affect that flow of workers. Finally, some of the funds that are expected to be made available by the water transfer agreement and state/federal grants to mitigate adverse socioeconomic effects will be usable to alleviate harm from temporary job disruption; these dollars could flow into the Imperial Valley economy.

Regarding the Salton Sea, some say the side effects are more real to outsiders than to the people of Imperial Valley, many of whom have neither used nor paid much attention to the Salton Sea. The valley is a fast-growing area with an expanding and diversifying economy. Frequently called a "breadbasket for the nation," it produces more than $1 billion worth of beef, milk, and vegetables every year. Major national retailers have moved in and are expanding. New beef-packing and cheese-making plants have brought jobs. Constantly building tract homes flank the cattle pastures and agricultural canals. The winter home for the famous Blue Angels of the U.S. Navy, Imperial County also has the large El Centro Naval Air Station. And a huge new El Centro shopping center like nothing else in the valley is viewed as a harbinger of better times.

In short, increases or decreases in Salton Sea recreational activities would not control Imperial Valley's economy. However, the growth in the economy has not occurred in the northern area of the valley around the Salton Sea. Wally Leimgruber, a director of the Salton Sea Authority, stressed that improvements in the quality of life at the sea would certainly be welcomed in the valley.

Those favoring the water transfer plan have consistently urged that the agreement could well lead to many hundreds of millions of dollars going to Imperial Valley for what are called "mitigation measures," including Salton Sea restoration. Indeed, MWD consistently emphasized those costs, predicting that they could soar into many hundreds of millions of dollars. Repeating his mantra that "it's all about money," MWD president Ron Gastelum consistently argued that MWD could ill afford to share such costs.

IMPORTANCE OF THE AGREEMENT

There is no doubt that the water transfer agreement has always been deemed vital to SDCWA plans for a "safe and reliable" supply of water for the San Diego area. The table on the following page projects both supply and demand for the years 2005 and 2015. (Note: Data is taken from tables in Chapter 9.) Again, MWD Supply is shown at the "firm" or preferential right amount on which SDCWA can conservatively rely.

Until 2015, SDCWA must depend on increased supplies of imported water from MWD in amounts far greater than its preferential right entitlement. Thereafter, that need decreases every year as the transfer arrangement water ratchets up to its maximum of 200,000 acre-feet annually.

SDCWA WATER SUPPLY-DEMAND PROJECTION
(ACRE-FEET OF WATER PER YEAR)

Year	2005	2015
DEMAND	697,326	760,795
Surface Water	85,600	85,600
Recycled	33,400	51,800
Groundwater	31,100	57,500
TOTAL LOCAL	150,100	194,900
Desalination	0	80,000
MWD Supply	344,800	345,895
IID Transfer	30,000	140,000
TOTAL IMPORTED	374,800	565,895
TOTAL SUPPLY	524,900	760,795
DEFICIT	172,426	0

From the beginning, it was recognized that numerous obstacles would have to be overcome before the water transfer agreement could be finalized. Various local, state and federal governmental agencies had to be satisfied. Hundreds of millions of dollars had to be found for Imperial Valley mitigation measures. The Salton Sea problems had to be solved. Some provisions of California's endangered species act had to be waived. And viewed as most troublesome was a lawsuit filed by IID challenging the legal right of the Secretary of the Interior to cut off its Colorado River water.

In that lawsuit, IID charged that the Secretary unlawfully reallocated water from IID to more populous urban areas in southern California for political reasons and cut IID's water deliveries to "strong-arm" IID into signing the pending Quantification Settlement Agreement. In its petition to be allowed to intervene in that case, MWD pointed out with some justice that "dividing up Colorado River water is a 'zero sum' game: IID can increase its share only at someone else's expense" and argued that MWD is that "someone else." MWD also contended that the IID lawsuit inevitably affected existing agreements with respect to Colorado River water. MWD asserted that the Secretary's action was in effect an effort to enforce the so-called "reasonable and beneficial use" requirement of western water law, a doctrine that is anathema to IID and other water districts which bristle at claims by California's urban areas that they have been for years wasting large amounts of water.

THE FUTURE FOR AG-URBAN TRANSFERS

Maureen Stapleton, SDCWA's General Manager and the driving force behind the water transfer agreement, has constantly stressed that it is critical to meet San Diego's need for a reliable, properly diversified water supply. To use today's sometimes overworked cliché, its proponents have argued that this and other such arrangements can be "win-win" deals not only for San Diego but also for other urban and agricultural areas that could welcome similar arrangements providing water for urban areas and money for the agricultural interests.

This does not mean that these ag-urban deals will be in the future easy to make and open-ended in quantity. Only a finite amount of water can be taken from agriculture before there is an adverse impact on ways of life and a consequent public reaction. Witness a letter written to the *Imperial Valley Press* after completion of the San Diego ag-urban transfer deal. The letter suggested the local fear that after the "urban interests had established a beachhead in the Imperial Valley, their next move will be to construct a siphon." The writer criticized the view of some in San Diego that "one size fits all" with respect to environmental mitigation and warned against approaches that "taunt the valley or offend local sensibilities."

There are in other areas many possible "third party" impacts akin to the Salton Sea and job-loss issues that came close to derailing the Imperial Valley agreement. In particular areas, as in the Sacramento-San Joaquin delta, there are powerful environmentalists who can significantly complicate matters. There are also the problems and costs of transporting the water south from the Central Valley, the area offering the greatest potential for such transfers. While the north-south California Aqueduct is currently only about half full, that unused capacity can disappear over time. Some skeptics argue that all of these roadblocks may make urban-agricultural water transfers a transitional phenomenon.

However, the potential is too huge to ignore and the consequences of not pursuing agricultural transfers are too detrimental to ignore them. It makes little sense for precious water to be wasted when the demand in urban areas is so heavy. Inherent in much of western water law is a doctrine held high in an arid land that no one has a right to waste water and use it inefficiently. "Reasonable and beneficial use" is the phrase used to describe this critical doctrine that has become anathema to agricultural interests desperate to protect the water that grows the crops. This doctrine has considerable appeal to people in urban areas and their representatives in government. While ag-urban transfers will encounter both rising costs and mounting opposition from agricultural-related interests and an emotionally aroused public, their growth seems ordained by growing California populations demanding more and more water in the dry years that seem to lie ahead.

CHAPTER 11
JOURNEY TO AN
EPOCHAL AGREEMENT

(Author's note: Unless otherwise stated, quotations in this chapter were obtained during personal interviews conducted during 2003.)

When the year 2003 began, those governing SDCWA felt reasonably confident that it had developed an overall water management plan for San Diego County that would assure a safe and reliable supply of water for those who live and work in the county. Nevertheless, a measure of unease continued to exist that was directly tied to a deteriorating relationship between SDCWA and MWD. That unease grew stronger as SDCWA and MWD disagreed over the SDCWA plan to achieve water security through the water transfer agreement with Imperial Irrigation District. Little did anyone know in January, 2003, that this effort would lead to an epochal Colorado River water rights agreement that would command national attention. The beginning place in describing the tortuous journey to that epochal agreement is the relationship between SDCWA and MWD.

The decades-old dichotomy between what have been roughly termed the "regional" and "independence" approaches on critical water supply issues was described in Chapter 6. The former approach argued that reaching decisions should be constantly and firmly based on a consensus concept that "All of us in southern California are in this together and must find solutions together under the MWD umbrella." This concept was perhaps best expressed by Harry Griffen of the Helix Water District during the years he was called "Mr. Water" as he served on the boards of both SDCWA and MWD. While agreeing that diversification of supply is highly desirable for San Diego, he argued strongly in his *Harry Griffen Manuscript* that alternatives to achieve it with respect to imported water should be found by SDCWA working not separately but with and through MWD.

The second approach (here called "independence" for short) accepted the basic fact that San Diego must always rely to some extent on MWD's imported water but emphasized greater individuality and more freedom of action on water supply issues. In a very real sense, this approach began with a trio of women at SDCWA reacting individually (over a period of years) with dissatisfaction over MWD's handling of the severe 1987-1991 drought. The three women were Christine Frahm, a tough and smart lady who served on the boards of both SDCWA and MWD and briefly chaired the former; Francesca Krauel, another outspoken member of both boards who served as SDCWA chair for two years; and Maureen Stapleton, who moved from deputy city manager of San Diego to general manager of SDCWA. A strong, focused leader whom some initially called "Ice Lady" with respect for her determination, Stapleton took over a strong and capable staff largely built by her predecessor, Lester Snow.

Rightly or wrongly, San Diego became fearful in the 1990s of MWD's perceived unwillingness to recognize its water needs in a time of shortage. New to the MWD board, Frahm and Krauel adopted what some have termed an "in your face" opposition posture. Fairly or unfairly, depending on viewpoint, observers concluded that their actions made achieving consensus more diffi-cult and contributed to a deterioration of the working relationship between MWD and SDCWA. At SDCWA, Frahm worked closely with other board members and Stapleton in moving that body away from the regional approach toward a unilateral search for additional water supply sources that the staff deemed were essential and that would obviously lessen San Diego's dependence on MWD. As a result of this and the underlying issues previously discussed, the two agen-cies moved more into a confrontational posture. Supporting Stapleton as the years went by were other strong members of the SDCWA board, including Mike Madigan and George Loveland from the City of San Diego and Harold Ball from the Helix Water District.

Not all board members were happy with the new direction. Mike Leach admired the determination of Maureen Stapleton but still favored a consensus-seeking approach on difficult water issues that was predicated on a regional view. Leach served on the SDCWA board from 1986 to 2001, representing the City of San Diego in all but one of those years, and on the MWD board for approximately one year. He served briefly as chair of SDCWA in 1993-1994. Despite his misgivings about the movement away from constant collaboration with MWD and the economics of some diversification proposals, Leach joined others on the board in supporting Stapleton when she took the lead on the water transfer program involving the purchase of agricultural water from Imperial Valley.

MWD changed somewhat in the 1990s as, like SDCWA, it began looking even harder for water supply alternatives, and became increasingly concerned about SDCWA's policy direction. Leaders at MWD feared that SDCWA's ac-tions reducing water purchases from MWD could harm MWD's financial situ-ation. MWD had a very large, expensive staff, plus increasing capital costs related to infrastructure needs, and the revenue stream to cover those costs was not always consistent. While the City of Los Angeles did in some years in-crease its purchases of imported water from MWD, thus providing more revenue for MWD, it was basically a "roll on, roll off" buyer that did not consistently purchase over the years the same large amounts of water. Thus, the combination of Los Angeles' uneven purchasing and the possibility of San Diego obtaining other water supply sources could seriously affect MWD's steady supply of water sales income.

These financial problems confronting MWD were exacerbated by the constant need to cover capital costs through bond issues. General obligation bonds supported by property tax revenues were increasingly being replaced by revenue bonds supported by income from the sale of water. This meant that water prices to customer agencies had to be raised, which inevitably increased the importance of SDCWA as MWD's largest customer. Thus, MWD eyed with increased wariness the attempts by SDCWA under Stapleton's leadership to find alternative sources for San Diego's water needs. And, of course, MWD leadership continued to murmur that San Diego was being ungrateful, forgetting all the money that MWD and Los Angeles had provided to build infrastructure for the San Diego system. Add to these factors the two agencies' past conflicts that refused to go away and the result was continued confrontation.

PREFERENTIAL RIGHTS AND VOTING POWERS

A continuing sore spot has been disagreement over preferential rights to water in times of shortage and the related dispute over MWD board voting rights. San Diego continues to stress the imbalance between its water purchasing financial support for MWD's capital and operating costs and its "preferential right" to amounts of imported water from MWD in a time of shortage. Currently, San Diego has a preferential right to only about 15% of MWD's water, but regularly purchases a substantially higher portion, more than a fourth, of MWD's water. And San Diego's voting rights on the MWD board continue to be far less than those held by the City of Los Angeles. Although the purchasing disparity has ameliorated somewhat as the City of Los Angeles has in some years increased its purchases from MWD, the disparity is still considerable.

SDCWA finally decided to bring the preferential rights issue to a head by filing a lawsuit alleging that it was not the original intent of the drafters of the troublesome Section 135 to demand a formula that excluded purchase money for water from MWD's "income pot" that determined both voting and water rights. MWD clearly resented the lawsuit. Philip Pace, MWD Chairman, charged in a written statement that SDCWA constantly "deludes itself" about water issues, preferring constantly to just "blame Los Angeles." San Diego has never forgotten that in 1984, City of Los Angeles representatives on the MWD board sought to overturn the Laguna Declaration's command that MWD meet all members' water needs. Pace dismissed the SDCWA lawsuit as a foolish attempt "to get paper rights [not real water]" and charged SDCWA with leading an "amnesiac crusade" based on the 1990s drought-years controversy. To this day, MWD calls preferential rights a "phantom issue." Others have challenged SDCWA motives, saying that the lawsuit was more a "flag to wave."

Experienced San Diego water attorney Paul Engstrand has repeatedly echoed those who feel that San Diego has unduly emphasized the Section 135 issue. He agrees with MWD and some other local water officials that this is a "bogeyman issue," that Section 135 has never been and probably never will be invoked to control MWD water distribution. Board member George Parker disagrees: "Even though Los Angeles and Metropolitan say, 'Don't worry about it, it's going to rain,' I think we need to plan for the day when it's not going to rain." And even Paul Engstrand agrees in the final analysis that Section 135 should be rewritten because it is such a "thorn in San Diego's side," conceding that it is "arcane and out-of-step with how water shortages should be handled," and that the failure of one attempt to rewrite Section 135 should not foredoom other attempts.

Not surprisingly, President Gastelum of MWD simply brushed aside the Section 135 lawsuit, stating: "It's not going anywhere so why should I be unhappy about it?" The Superior Court has dismissed the lawsuit; it now rests on appeal to the California Supreme Court.

SDCWA's comprehensive 2000 Urban Water Management Plan for the future discusses other serious differences with its current MWD relationship, including particularly the voting rights formula. "The Authority believes," states the plan, "that Metropolitan's governance and voting structure should be changed to reflect the interests of those member agencies who are paying the bills." In other words, San Diego, as the biggest water purchaser from MWD, should have more votes on the MWD board. The plan states that "member agencies should get what they pay for and pay for what they get."

Gastelum of MWD does not attach great importance to this issue, arguing that the voting power of the City of Los Angeles has never given it undue control over MWD's policies. When pressed on the voting power disparity that nevertheless continues to exist, Gastelum perhaps understandably avoided the issue by taking refuge in a statement that it is his obligation to represent all the member agencies of MWD, including both San Diego and Los Angeles.

PROPOSED NEW CONVEYANCES FOR IMPORTED WATER

Being given serious consideration by SDCWA are two new proposals for bringing imported water to San Diego, both of which make MWD extremely unhappy. The first is Pipeline No. 6 to bring water to San Diego that would originate at MWD's Lake Skinner treatment plant complex in Riverside County. The northern portion of the pipeline from Lake Skinner to San Diego County would have to be built by MWD. The remainder of the pipeline would be built by SDCWA with a tunnel through Mount Olympus and then south to the current Twin Oaks Valley facility. SDCWA currently gives this project a low priority, concluding that it first wants to fully explore desalination. If that is successful, there would be no need for Pipeline No. 6.

The second proposal is a canal that could bring Colorado River water directly to San Diego and possibly also to Mexico's northern Baja California which needs the water. Some call it a bi-national canal but SDCWA has named it the Regional Colorado River Conveyance, stressing that it need not run into Mexico. Engineering feasibility studies have been completed on different routes from the Colorado River to Imperial Valley, all utilizing as much as possible of the existing All-American Canal. Only one of these routes goes south of the border to make Mexico a necessary partner.

This proposed canal could supply Colorado River water to Mexico and also could convey the water coming to San Diego from Imperial Valley under the pending water transfer agreement. As presently planned, that water would be conveyed to MWD's Colorado River Aqueduct and thence to San Diego, with SDCWA paying a "wheeling charge" to MWD. If the proposed canal were built, SDCWA could undertake the massive task of building a new aqueduct to bring the water to San Diego over the mountains.

To those who remember history, this situation is eerily similar to the choice that San Diego faced at the end of World War II when it decided to take its share of the Colorado River water from MWD's Colorado River Aqueduct rather than going over the mountains to hook-up with the All-American Canal in Imperial Valley. SDCWA has completed feasibility studies for the mountain traverse. One route would lift the water 3800 feet over the Peninsular Ranges and the other would use a 34-mile tunnel through the mountains, requiring a lift of only 1500 feet. Both routes would bring the water to San Vicente Reservoir. Total cost of the proposed new canal together with the conveyance through the mountains is estimated by SDCWA at $2.6 billion.

MWD is opposed to the project. If only the Imperial Valley segment were built, MWD would lose less since it would recoup some of the lost SDCWA purchase money through the wheeling charges it would collect for conveying the water to San Diego. But if the project continued through the mountains, MWD would not obtain the income it would receive if the water was conveyed through its Colorado River Aqueduct. MWD will also argue, with justice, that the money for this project could be better spent elsewhere.

SDCWA is far from committed to either Pipeline No. 6 or any part of the Regional Colorado River Conveyance. As to the latter, Mike Madigan, City of San Diego representative on the SDCWA board, has ambivalent feelings but he likes the precedent for cross-border cooperation: "If we do something with water, could we do something with a railroad?" On the other hand, Madigan cautions that it must make economic sense. All agree that the project would take a long time to complete. One old hand at SDCWA commented that by the time it could be brought into operation, "there may be desalination that would make the project as obsolete as a Roman aqueduct or the dodo bird."

A $1.7 million study funded by the U.S. Environmental Protection Agency has recently recommended that San Diego and Tijuana focus on wastewater reuse and desalination. "The bi-national canal perhaps isn't quite the deal it was a few years ago," concludes Madigan, adding "desalination is moving a little more quickly."

CONTINUED DISPUTE OVER THE
WATER TRANSFER AGREEMENT

From its inception, there was controversy between MWD and SDCWA over the Imperial Valley water transfer agreement and MWD was charged repeatedly with taking action to jeopardize the agreement while at the same time continuing to protest that it has long favored finding ways to bring wasted agricultural water to California's urban areas and has itself negotiated several agreements to do that.

In February 2003, MWD filed a formal "confidential" memorandum with the California Department of Water Resources. Leaked to the media almost immediately, the gloomy memorandum signed by Philip Pace, chairman of the MWD board, asserted that MWD has "long supported" the water transfer but questioned the high level of state funding for the plan and complained that it is "too tilted" toward San Diego. In addition to this, while negotiations for San Diego's water transfer agreement were still underway, MWD began questioning San Diego's need for the water. "There is no water crisis," said MWD board chairman Philip Pace. President Gastelum put it a little differently, quoted by Michael Gardner in the *San Diego Union-Tribune* as saying, "We have been consistently saying the sky does not fall if this deal doesn't come through." But then he upped the ante.

In an interview on *The CBS Evening News*, Gastelum made the flat statement, "We have a 20-year supply of water in southern California." A follow-up clip showed the manager of an Orange County facility drinking water reclaimed from local wastewater, sending the obvious message that San Diego should look to local water reclamation and other water conservation and local supply implementation measures.

San Diego Union-Tribune reporter Michael Gardner called Gastelum's statement a "stunning shift" by MWD that "could scuttle the water transfer agreement and sink the proposed seven-state accord on the Colorado River" (referring to the previously discussed Quantification Settlement Agreement). Gardner also noted that "the water transfer agreement would allow San Diego a significant new measure of independence from MWD."

San Diego officials were shocked but on reflection not surprised at the Gastelum statement. "Be realistic," said Harold Ball, who admits to initial concern, "After all, Colorado River water is free for MWD and they sell it to San Diego at a very high price. Why would they let San Diego go over to Imperial Valley to get cheaper water?" Then he added, "Folks on the SDCWA Board are convinced that Gastelum has been all along hoping and working for failure of the San Diego-IID deal because this would help keep MWD as the sole water provider for southern California." When asked his opinion as to why MWD chose to undercut the San Diego positions on water supply, George Loveland, San Diego City representative on the SDCWA board, responded: "Because of pressure put on them and because they don't like the IID water transfer deal. They want to be the water supplier for all of southern California."

Deputy General Manager Dennis Cushman reacted much the same way, agreeing that Gastelum "certainly undercut the San Diego message." He pointed out that "Two weeks earlier MWD was saying that we had only a two-year supply" and that anything less than a six-year supply makes the water-seekers very uneasy in view of the drought history of southern California. Cushman stated that part of MWD's motivation is to "keep their cash cow harnessed" and that their tactics reflect a degree of arrogance. "MWD doesn't seem to realize that we're on to their game," he said. But Cushman was more moderate. "They understand our water-search aims," he said politely, "they are just not happy about the way San Diego goes about it." And he repeated several times, "We don't seek independence from MWD; we're just seeking to diversify our water supplies."

When asked about the controversy surrounding his statement, MWD president Gastelum pointed out that MWD has a legal responsibility to answer queries about its capability of supplying needed water for up to 20 years and also asserted that he was doing nothing more than responding to such inquiries. California law does mandate that cities and counties must verify that there is enough water before allowing major developments and it is true that major developments were pending in the area served by MWD when this controversy arose. However, some say that this might not justify volunteering a 20-year supply statement on a national television news program when one of the MWD member agencies was stressing the critical necessity for new water supplies.

Gastelum strongly denied that MWD had opposed or undercut San Diego in the negotiations concerning the water transfer agreement. Stressing that "it's all about money," he stated that MWD only wishes to see a proper balance in allocation of costs. "San Diego," he said, "too often likes for others to pay for infrastructure and is deathly afraid of assuming such costs." He cited as an example the controversy over wheeling charges for conveying the Imperial Valley water to San Diego through the Colorado River Aqueduct that was financed and built by MWD. Gastelum also cited his concern about allocations of money from Proposition 50, the state water bond measure recently approved by California's voters. Gastelum mentioned specifically the sought-for use of that money to mitigate the Imperial Valley environmental and socioeconomic consequences of San Diego's water transfer agreement.

IS INDEPENDENCE FROM MWD
A SAN DIEGO IMPERATIVE?

As previously discussed, the current San Diego policy regarding the search for new water supplies has been called by some a "search for independence" on water supply whereas others, including General Manager Maureen Stapleton of SDCWA, term it less challengingly as a "search for diversification" of water supply. Philip Pryde, professor emeritus at San Diego State University, looked at this "independence issue" twenty years ago when he was asked by the San Diego County Board of Supervisors to help head a task force on the question of whether independence for San Diego on water supply was feasible. His and the task force's conclusion? To paraphrase, "Sure, if you want to pay for it."

There is no question that the water transfer agreement issue aggravated the relationship between SDCWA and MWD, already strained by other difficult, sometimes emotional issues. Some said during the disagreements over the agreement that MWD board members became increasingly hostile to San Diego, although MWD's Ron Gastelum has disliked that word, contenting himself with saying that "they just don't understand what SDCWA is doing." Some at SDCWA have not been so polite in their assessments, concluding that MWD has acted like an "800 lb. gorilla" hungry for southern California water supply control. Continuing the search for mammalian metaphors, Dr. Pryde was asked whether MWD could be called "the elephant in San Diego's living room." With a smile, he replied, "I agree but would add that much depends on how well the elephant is housebroken."

Have all the episodes suggesting an MWD propensity to quell efforts by San Diego to control its own water supply destiny caused SDCWA to search for independence from MWD? Maureen Stapleton, General Manager for SDCWA, flatly refuses to use the word "independence." Those who support her state that she cannot conceive of any set of circumstances in which San Diego would give up that considerable supply of water. Expressly denying the charge by some observers that she is "on an independence kick," Stapleton persists in the somewhat euphemistic mantra voiced by all of SDCWA General Counsel Dan Hentschke, Senior Deputy Dennis Cushman, and George Loveland, Senior Deputy San Diego City Manager, who is a representative on both the SDCWA and MWD boards: "SDCWA is merely seeking diversification of supply." None of them denies, however, that every gallon of water obtained through diversification is one less gallon that has to be bought from MWD. Is this a search for independence? Is the glass half full or half empty?

Many of these officials use their preferred metaphors: "We mustn't put all our eggs in one basket" and "we cannot hang our hat on one solution" before concluding unanimously that "the solution for our problems is diversification." Insisting that there is a realistic difference between this statement and a "craving for independence," Cushman agreed that he and Stapleton sedulously avoid the "i word," repeating that "we just want to control our own fate." All appear reluctant to overtly criticize MWD but when pressed, Cushman did agree with the earlier-quoted statement by Harold Ball: "MWD wants to be the water supplier for all of southern California and we find that desire to be very troublesome."

Dan Hentschke, SDCWA's General Counsel, addressed the issue in a different manner: "We do work together on all fronts; the other view is emotional and rhetoric. Ninety-eight per cent of the time we work in cooperation. All we want is balance." But even this attorney conceded that "the other two per cent can be troublesome." While he agreed that MWD does appear at times to have a desire to control southern California's water supply, he would not say that MWD thinks it knows what's better for San Diego than San Diego does. However, there are other officials at SDCWA who would make that strong statement, although they prefer to remain anonymous.

The traditional way that San Diego has met its continuing deficit between demand and supply has been to rely on MWD to supply the needed water. Or, as SDCWA states in its current Urban Water Management Plan, the reliability of San Diego's water supply "is subject to the discretion of the Metropolitan Board of Directors."

A clear majority of the SDCWA board is uncomfortable with this situation and supports Stapleton's strong policy of seeking further water supply diversification. The vote in favor is perhaps as high as 80%-90% if weighted votes are counted and still well over a majority on a member agency head-count basis. A minority, mostly North County agencies, have favored more reliance on MWD for water supply decisions. These seven "rebel" districts have called themselves the "Economic Study Group." A few have stated openly that they wish they could be members of MWD rather than SDCWA, feeling closer to the former. The City of San Diego, perhaps dominant but certainly not controlling on the SDCWA board and led by its representatives Mike Madigan and George Loveland, with support from Harold Ball of the Helix Water District, clearly support Maureen Stapleton on what she and others in command at SDCWA call the "diversification of supply" policy.

LITIGATION POSITIONS

There are some revealing sections in SDCWA's legal brief filed in the previously discussed preferential rights litigation that have importance beyond the issues in that case. The brief cites a litany of anti-San Diego actions by MWD and the City of Los Angeles (a separately named party to the action), stressing San Diego's "forbearance" for years from seeking water supply alternatives followed by MWD "affirmatively acting to thwart" San Diego when it sought the Imperial Valley agricultural water.

The SDCWA brief concludes the litany with strong language: "The 'forked-tongue' approach of MWD and Los Angeles has created uncertainty that is unreasonable and intolerable under the circumstances." Few of the specific differences cited by San Diego are expressly contradicted in the responding briefs filed by MWD and Los Angeles. Each is largely content to assert legal arguments based primarily on the statutory language of Section 135 that appears to preclude inclusion of water purchases in determining water supply and voting rights. However, the brief filed by Los Angeles goes further, arguing that San Diego's position on this issue "would encourage consumption of water, which is not desirable in an arid environment such as the West." San Diego attorneys do not like this statement since it ignores the tough water conservation measures adopted by SDCWA and its member agencies for the last ten years.

The Los Angeles brief does note San Diego's reference to a past agreement by MWD and Los Angeles that the present water preference system "constitutes an inequity," but then goes on to state dismissively, "it does not matter" (p. 16). Repeatedly, Los Angeles asserts (e.g., p. 19) "San Diego perhaps has never liked the Section or the impact it has on its future ability to provide water to its constituents, but it has always known of it." There is no answer to the counter that so has MWD, having been constantly reminded of the inequities, but never endeavored to remedy them.

MWD argues that San Diego should be content to rely on the powers given by law to MWD for dealing with emergency shortages of water, stressing that MWD can exercise its "sound discretion on that hypothetical day (should it ever dawn)" to supply additional water. Then MWD castigates SDCWA for telling MWD "how to exercise its discretion upon the happening of an event that has never occurred under a statute, Sec. 135, that, should the event occur, would not control its discretion anyway." These confusing words can be read to say that MWD makes no firm promises, no commitments with respect to allocations in a time of critical water shortage.

TWO SIDES TO THE DISPUTES

The City of Los Angeles and MWD continue to make the argument that it was Los Angeles and MWD, through leadership, property tax monies, and bond issues, that brought Colorado River water to southern California and that San Diego was a prime beneficiary of that investment. And it was the same money sources that financed the aqueduct and pipeline bringing that water south to San Diego. The conclusion: Since MWD and Los Angeles have in part subsidized San Diego's access to and utilization of Colorado River water, SDCWA is in a poor position to complain about paying more in recent years toward MWD's capital and operations costs. And Los Angeles representatives on the MWD board would be quick to point out that in recent years it has been paying more to MWD through increased purchases of imported water.

Steve Erie, the water-knowledgeable UCSD professor, argues strongly that Los Angeles did "all the heavy lifting" to make Colorado River water possible for San Diego. He, along with MWD president Ron Gastelum, finds merit in cost-sharing criticisms of some of SDCWA's activities with respect to water supply. "Face it," says Erie, "San Diego has often been a free rider on infrastructure projects paid for largely by Los Angeles." Gastelum agrees, saying "San Diego looks for others to pay for infrastructure."

However, and despite these contentions, SDCWA still maintains that it finds itself in a frustrating position, constantly referring to the legal fact that its imported water supply from MWD is always subject to change at the discretion of the MWD board. The current Urban Water Management Plan does recognize that MWD's existing Water Surplus and Drought Management Plan contains a "contingency analysis" describing the actions MWD could take in the event of water shortages affecting member agencies. At best, though, SDCWA states, this "leaves the SDCWA position in a shortage situation uncertain."

Even stronger is the SDCWA Plan's conclusion: "However, the Authority recognizes that Board actions at Metropolitan could change the terms of the WSDM Plan at anytime and therefore the WSDM cannot be relied upon to ensure the reliability of Authority supplies."

John Economides, Chief Engineer for SDCWA who once worked for MWD as an assistant engineer, stated his support for the current SDCWA policy this way: "Metropolitan has statutory responsibility for distribution of imported water throughout its domain. But being in overall charge does not mean total control and Metropolitan will not leave us alone as we seek to diversify our supplies. Metropolitan seeks to control all activities of its member agencies that may have an effect on usage of imported water. That's reaching too far." In other words, the quarrel is not with MWD's mission; it is with the manner in which it executes the mission.

Economides agreed to some extent with those who support the so-called regional approach to water issues by stating that all of southern California would be better off if SDCWA and MWD could find ways to work in concert more than they do at present. For example, he expressed his belief that MWD should have been saying that San Diego's water transfer agreement was a good thing. Instead, he and others perceive that MWD was not only opposed to the agreement but undermined it because they did not control it. This was "unconscionable," said Economides. One of his key assistants, Kenneth Steele, characterized this as part of the MWD determination to "put San Diego down."

By no means all of SDCWA's present and past officials are negative about MWD and even those who criticize it are quick to say that there have been and are many areas where the relationship has been fruitful. As earlier noted, former SDCWA attorney and water-war veteran Paul Engstrand had praise for MWD. He emphasized that MWD's policies and actions have "for more than 50 years resulted in it supplying a reliable supply of imported water" and decried what he called "the present belligerency toward Metropolitan that seems rooted in misconceptions about Metropolitan's legally structured role, personality clashes, and Section 135."

Engstrand was not impressed by San Diego's proposed water transfer agreement hailed as progress toward diversification of supply. "It is more like changing brokers than diversification," he asserted, and concluded overall that "those who favor the proven regional approach would encourage continued support for Metropolitan's announced plans for the future." Lin Burzell, another old-timer, said, "I just don't believe that MWD is the bad guy. I never felt there was a problem with MWD that could not be resolved. We've often got just what we wanted from them; repeatedly, they have come around to our point of view." Burzell then cited specific examples where this has happened in the past. Taking up the water transfer deal, he stated, "The best way for San Diego to go is to have MWD represent us."

Engstrand would still like to see San Diego work more within the MWD board to achieve consensus, as he said it has in the past, on measures that would bring adequate and reliable supplies of water to all areas of southern California. He praised MWD's initiation of the Laguna Declaration that water would be supplied according to need and MWD's 2000 Regional Urban Water Management Plan mission statement as a strategy that would provide the region with a reliable and affordable water supply for the next 25 years. "In consensus building," says Engstrand, "there are bound to be some disagreements but reliance on Metropolitan fulfilling its mission statement is fully justified." He concluded, "San Diego's energies should be spent on developing consensus for sound strategies at Metropolitan instead of wasted on the unattainable independence posture."

Dale Mason, who served for years on the MWD board and has represented Vallecitos Water District on the SDCWA board, joined Engstrand and Burzell in supporting the regional approach. While in the final analysis he supported the Imperial Valley water transfer agreement, he called it a "transfer of money, not water, and continues to feel that it unduly "punishes" MWD because it would substantially lessen SDCWA's purchases of Colorado River water from MWD. Overall, Mason joined Engstrand in arguing for a consensus approach within MWD. "Instead of working to build a coalition within MWD," he argued,

"SDCWA has isolated itself for a long period of time, going out of its way to reduce water purchases from MWD by finding alternative water sources."

FAILURE OF THE QUANTIFICATION SETTLEMENT AGREEMENT

The most disturbing action by MWD during the water transfer agreement dispute was its refusal to meet the Department of the Interior deadline of December 31, 2002, for execution of the essential Quantification Settlement Agreement (QSA) that was negotiated by all recipients of the Colorado River water in southern California. That agreement, as previously explained, was designed to provide a "soft landing" for California to reduce its river water usage to the 4.4 million acre-feet annually required by law and other existing agreements.

When the deadline for QSA execution was not met, California's annual water entitlement automatically went down to the 4.4 maf level. This meant that MWD's entitlement would fall from 1,250,000 to 713,500 acre-feet, a 43% reduction from its anticipated need. And the situation was made worse by the IID request that the Secretary of the Interior reduce MWD's Colorado River entitlement by another 192,000 acre-feet. Altogether, MWD faced a cut of approximately 728,500 acre-feet, a 58% reduction from its then-existing entitlement.

Publicly, MWD played down the results of its rejection of the QSA and the consequent substantial reduction in MWD's Colorado River water supply. On March 25, 2003, MWD sent its *Report on Metropolitan's Water Supplies* to its member agencies. The report stated that the absence of an executed QSA is "serious," however "it is not an emergency." Reviewing its extensive water storage facilities and a series of enhanced conservation and local supply development programs, the MWD report stated that it could meet all member agency demands over the next 15 to 20 years, even taking into account the ordered reduction in supply from the Colorado River.

Asked by the SDCWA board to respond, its staff made an extensive analysis that ended up expressing considerable skepticism about the report's statement that MWD could supply member agency needs for 15-20 years. The SDCWA staff critique discussed at length many questions about the accuracy of both the report's data and its optimistic projections, concluding that MWD had by its actions bringing about the failure of the QSA placed its member agencies "at significant risk."

In July, 2003, MWD filed a formal document in the pending litigation with IID that was in sharp contrast with MWD's reassurances to member agencies that it could meet all water demands for the next 15-20 years without the reductions in Colorado River water supply resulting from the QSA failure. The document, entitled "Declaration," was a statement of facts under oath, in affidavit form. The Declaration opened by stating that the reductions "represent a significant and immediate impact to MWD's water supply plans, affecting both water supply and water quality for the region."

The Declaration described at length the serious consequences to MWD of the reductions. Featured was the assertion that "MWD will have to resort to drawing upon regional storage reserves." These, as the Declaration explained, constitute "vital supplies which had been earmarked for drought and emergency storage (in the event regional supply lines were lost due to earthquake or other catastrophe such as terrorist attack), potentially compromising regional water supply reliability in future years for the nearly 18 million southern California residents in Metropolitan's service area."

These dire consequences, none of which were contained in the previously described report MWD sent to its member agencies, are repeated for emphasis in the sworn MWD Declaration: "To the extent that these storage reserves are reduced to offset a substantial reduction in Colorado River supplies for 2003, the ability of the region to survive multi-year droughts and to implement water management programs is compromised." The Declaration went on to state that the reduction in supply "will manifest itself as real hardships to MWD and its service area. There are no other alternatives available to prevent or further minimize these hardships."

Making even sharper the contrast between its optimistic reassurances to member agencies and its pessimistic outlook filed under oath with the court, the MWD Declaration stated that the "significant reduction," ordered by the Department of the Interior, "increases the risk that MWD will not be able to meet Federal and State drinking water quality standards this year."

Strangely, MWD president Gastelum ignored his agency's affidavit when he later spoke publicly about the region's water problems. In a column in the *San Diego Union-Tribune*, July 10, 2003, columnist Logan Jenkins reported on a Gastelum appearance at a meeting of the Vallecitos Water District, a member agency of SDCWA. In his column, Jenkins reported the current MWD message that "the Met will be awash in water, even if the Colorado dries up as a source," a message strongly disputed by SDCWA, fully aware that the "Met message" had undercut the chances for finalizing the then-pending and much-desired IID water transfer agreement.

At the end of Gastelum's presentation at the Vallecitos meeting, he was confronted by David Fogerson, a SDCWA engineer. Fogerson said that Gastelum's presentation was squarely in conflict with the Declaration's description of "dire consequences" from the lack of Colorado River water. Waving the document that MWD had filed under oath, Fogerson said, "These are Met's own words." Logan Jenkins reported Gastelum's reaction: "Gastelum waved his hand as if to forgive such charming naivete. 'Let's look at the wet water and not look at semantics in litigation,' he gently chided."

Jenkins also wrote that, to MWD, "'semantics' evidently means 'lies.' One wonders about the reaction of the court if it had known that MWD was publicly characterizing its own affidavit under oath as meaningless "semantics."

"Wet water" is old-timers' water talk that means "real water" as opposed to planners' projections, hopes, and estimates. In other words, "what you can really count on," "what you can take to the bank," to use another metaphor.

The episode vividly depicts the double track that MWD followed during those months: undercutting the need for San Diego's water transfer agreement by saying that a plentitude of what its president facetiously called "wet water" was already available in southern California, while arguing in court that MWD desperately needed more Colorado River water.

MWD's actions, commencing with its rejection of the QSA and culminating with its contradictory statements, puzzled members of the SDCWA board. They felt strongly that the actions jeopardized San Diego's pending IID water transfer agreement. SDCWA was deeply concerned about its ability to meet future water demands without the water transfer arrangement, as the following table shows. (Note: Data is taken from tables in Chapters 9 and 10.) MWD supply is computed on the basis of San Diego's preferential rights.

SDCWA WATER SUPPLY-DEMAND PROJECTION (ACRE-FEET OF WATER PER YEAR)		
Year	2005	2015
DEMAND	697,326	760,795
Surface Water	85,600	85,600
Recycled	33,400	51,800
Groundwater	31,100	57,500
TOTAL LOCAL	150,100	194,900
Desalination	0	80,000
MWD Supply	344,800	345,895
TOTAL SUPPLY	494,900	620,795
DEFICIT	202,426	140,000

The water transfer agreement was projected to supply 30,000 acre-feet in 2005, going up to 140,000 in 2015. Failure of the water transfer agreement would have wiped out San Diego's chances of eliminating that 2015 deficit.

Opposition to San Diego's water transfer agreement could be read between the lines of a letter written on June 3, 2003 by MWD Board Chairman Philip Pace to Governor Gray Davis. The letter highlighted what it called "threats" by the governor against parties that failed to agree to the QSA as required by the transfer agreement.

Pace's letter was fully informed by the repeated refrain of his president, Ron Gastelum, that the water crises of southern California are "all about money." The main thrust of the letter was opposition to "diversion of Proposition 50's funds to pay a significant portion of environmental mitigation costs of the market transfer component [that's San Diego's water transfer deal] of the QSA." The letter argued that this would violate the "voter mandate" that Proposition 50 funds be "made for specific purposes, not including subsidies for market transfers." The letter also stated that the agreement "may be flawed in the absence of a definitive determination on restoration of the Salton Sea," a statement that would definitely have been welcomed by Imperial Valley opponents of the water transfer agreement.

Complicating the picture was MWD's request to the state to use Lake Oroville for storage of 100,000 acre-feet of water it would gain from its Sacramento water transfer agreement. Governor Davis refused permission and this was viewed by some as state pressure on MWD to go along with the San Diego water transfer agreement. In a counter move, the Interior Department appeared to side on this issue with MWD when it informed MWD that it could store the water at Lake Shasta north of Redding. In its letter, the Department of the Interior warned that allowing water from agricultural-urban transfers to escape into the ocean under the Golden Gate Bridge would make it difficult for other western states to join in efforts to secure more water for California from the drought-stricken Colorado River basin.

The QSA issue remained undecided until the very end of the dispute over the San Diego water transfer agreement. Observers have questioned why MWD chose to throw that monkey wrench into the already complicated southern California water supply machinery. One suggestion was that MWD was endeavoring to dramatically demonstrate the control it asserts over imported water supplies for all of southern California. In other words, said one observer, the 800-pound gorilla was beating its chest for all to hear and fear.

The failure of SDCWA, IID, and MWD to finalize what came to be known as the San Diego ag-urban water transfer agreement caused growing concern during 2003. Northern California legislators angrily accused MWD of trying to kill existing Colorado River agreements in order to prevent San Diego, its biggest customer, from throwing off its water dependence on MWD. MWD was criticized for making a deal to obtain water from northern California rice farmers, reopening old sores about taking water from the delta, while it sought to scuttle the San Diego agreement. Said the *Los Angeles Times*: "A growing number of politicians, environmentalists, and newspapers in southern California fear that the MWD will turn northward for more water." Specifically mentioned was pressure to increase the pumping of more water out of the Sacramento-San Joaquin delta through the California Aqueduct down to southern California.

THE CRITICAL SACRAMENTO NEGOTIATIONS

Finally, in late August 2003, representatives of SDCWA, MWD, IID, and Coachella Valley Water District met in Sacramento with representatives of Governor Gray Davis and the U.S. Department of the Interior, seeking ways to resolve the festering differences over the San Diego agreement. At the outset of the negotiations, MWD repeated that there is ample water for southern California for the next 20 years. However, as the negotiations intensified, considerable progress was made on the difficult issues, tentative agreements were reached, and MWD even indicated that it could put $82 million "in the pot" for Imperial Valley environmental mitigation measures. This encouraged the parties to believe that they were on the verge of consummating a deal that would "bring water peace to the West."

Then, on August 21, suddenly and amazingly to the hard-working negotiators, MWD retracted its $82 million gambit and rejected the tentative accords that had been painfully reached. Then, to the open-mouthed dismay of the participants, MWD dramatically walked away from the bargaining table. Officials from the water agencies with the most at stake in the negotiations swiftly and publicly expressed their outrage, stressing how hard they had worked to come together on what looked like a breakthrough agreement and emphasizing that the deal MWD rejected met every objection that MWD itself had asserted earlier in the year when contesting a QSA settlement.

MWD was accused of "sabotaging" the talks by critics who harked back to MWD's earlier efforts to torpedo the San Diego-Imperial Valley ag-urban agreement. The Southern Nevada Water Authority charged that "Metropolitan's obsession with protecting its stranglehold over California water supplies will cost southern California tens of millions of acre-feet of reliable water, pit water agency against water agency, and potentially state against state."

Going further, the Nevada official added, "That one rogue agency can utterly upend nearly a decade of work to establish peace and reliability on the Colorado River is beyond comprehension." Other water agencies and government officials joined in the attacks on MWD. State Senator Dennis Hollingsworth, who had worked hard to bring the parties together, issued a strong statement charging that "MWD never intended to reach a peaceful agreement." MWD was warned that its aberrant actions could "impact the quality of life for all southern California" and lead to "distrust among our friends in northern California."

Recognizing that all this was happening while the California recall fight was building up, the *San Diego Union-Tribune* wrote editorially, "It's no coincidence that [MWD president] Gastelum and his staff are making mischief in the midst of California's political chaos." "The gorilla is on a rampage," said one knowledgeable observer close to the scene. SDCWA General Manager Maureen Stapleton said, "Although MWD apparently doesn't see value in the deal, everyone else does."

Perhaps most significantly, Coachella Valley Water District shelved for the first time its decades-long disputes with neighboring IID, joining both IID and SDCWA in a strong statement threatening that the three agencies would go their own way without MWD: "If MWD wants to be left out in the cold, so be it." Thus, all other major southern California water agencies dependent on Colorado River water (Coachella, Imperial, and SDCWA) were united against MWD.

Looking back later on, Dennis Cushman, Deputy General Manager of SDCWA, said it was this statement plus two other factors that brought MWD back to the bargaining table. The first factor was pressure from other Colorado River basin states who wanted "peace on the river." The second factor was pressure from both the State of California and the federal government, not just the Interior Department but clear word that the White House wanted a Quantification Settlement Agreement that would bring water peace in the West.

To put the matter rhetorically and dramatically, MWD had put itself "in the bull's-eye" as the target for everyone and was finally forced to recognize that if it did not cooperate to reach an overall Colorado River settlement, "the dead cat would be on MWD's doorstep." So, finally, MWD sent word to the other parties, "Let's talk."

THE FINAL LANDMARK AGREEMENT

When the parties began talking anew, MWD stressed that while it would resume negotiations, any monetary participation in funding environmental mitigation measurers for Imperial Valley was "off the table." When this position was reluctantly accepted by the other parties who had argued that MWD should pay its fair share among those taking agricultural water from the valley, the pieces came back together and all parties finally accepted what has rightly been called a "landmark agreement." Following is a list of the essential terms.

- SDCWA will receive up to 200,000 acre-feet per year of Imperial Valley water under its 1988 agreement, commencing with a trickle of 10,000 acre-feet in year one and escalating year by year thereafter.

- SDCWA will be responsible for lining the All-American and Coachella canals with the State of California obligated to pay the costs up to $235 million. In return, SDCWA will receive for 110 years 77,000 acre-feet per year from the All-American Canal.

- Up to $300 million will be made available for socioeconomic and environmental costs in Imperial Valley, including Salton Sea restoration.

- A new state-appointed board will oversee the Salton Sea restoration plan, presumably in cooperation with the existing Salton Sea Authority.

- IID will provide up to 1.6 million acre-feet of water to the state which in turn will sell it to MWD with proceeds dedicated for mitigation measures and Salton Sea restoration.

- None of IID, SDCWA, Coachella, and MWD will be liable for any further obligations to restoration of the Salton Sea. The State of California will hold these agencies harmless from any liabilities resulting from environmental mitigation measures.

- SDCWA will give up its existing available discount on the "wheeling charge" for moving its Imperial Valley ag-urban water through MWD's Colorado River Aqueduct.

- IID will determine how to create the Imperial Valley ag-urban water, beginning with the fallowing or idling of farmland and moving on to a series of conservation measures to be adopted by Imperial Valley farmers.

- IID will continue to transfer 104,000 acre-feet per year to MWD in confirmation of their 1988 agreement extended to the life of the QSA.

- The Quantification Settlement Agreement, as modified to incorporate new terms required by the water transfer agreement, will be executed by all necessary parties.

- IID will dismiss its existing lawsuit against the Department of the Interior and IID will be assured by the Department of the Interior of its share of Colorado River water.

SDCWA chairman Bernie Rhinerson immediately announced: "It is a historic deal for the residents of San Diego County." Other parties joined in the acclaim, but some Imperial Valley farmers objected. One radio station owner in Imperial Valley said, "I urge this board not to put the final nail in this valley's coffin."

SDCWA, MWD, and Coachella Valley Water District quickly held board meetings to approve the deal. Last to vote was the IID board but it finally went along on a close 3-2 vote. Then, the necessary implementation legislation was adopted in Sacramento. Finally, U.S. Secretary of the Interior Gale Norton appropriately chose Hoover Dam for the formal document-signing ceremony, calling it "one of the most far-reaching water deals in this nation's history."

MEANING AND PORTENT OF THE DEAL

Even anti-California hard-liners in the Colorado River basin states expressed "admiration and surprise" at the ending of what one official accurately called "a nine-year roller-coaster ride." Governor Davis proclaimed, "there is now peace along the Colorado River and California's water future is now sparkling." And informed observers recognized that the 1988 San Diego-Imperial Valley water deal had in retrospect been pivotal in bringing about what were called "watershed" and "epochal" agreements promising "nothing short of a new era for water in the West."

National attention underscored the importance of the settlement. Even the *New York Times* rated it as worth an editorial entitled "Living with Scarcity in California" but could not resist the opportunity to give our state a supercilious and unwarranted slap: "California now promises to end its larcenous ways," referring of course to the state's over-usage of Colorado River water. The libel is unwarranted since the water came from Colorado River surplus plus unused allocations to Nevada and Arizona and was allocated legally by the Department of the Interior. *The Christian Science Monitor* also gave Norton "a pat on the back" for the deal but missed the mark by concluding that it "means California farmers must work to reduce their water use by planting fewer crops or growing those that require less water." San Diego's *North County Times* noted that "more than 30 million acre-feet of water — enough to cover the state of Pennsylvania in a foot of water — will move from farms to cities over the 75-year life of the deal."

The overall deal has these three critical meanings:

- SDCWA has succeeded in obtaining what it started out to achieve: a meaningful measure of independence with a diversified water supply. And all the water coming to San Diego from Imperial Valley is Priority 3 water from the Colorado River, ahead of MWD's Priority 4.

- Following up on President Gastelum's adjuration that "it's all about money," MWD succeeded in forcing SDCWA to pay considerably more for transporting its Colorado River water through MWD's Colorado River Aqueduct while at the same time MWD avoided any up-front responsibility for funding Imperial Valley environmental mitigation measures. The up-front commitments are by the State of California, SDCWA, Coachella Valley Water District, and IID.

- Up to as much as $300 million will be available for Salton Sea restoration. A considerable portion of this will come from MWD purchases of water obtained by the state from IID under the settlement agreement.

The first of these is the most important for San Diego County since it involves the county's water supply future. To appreciate this, take 1991 as a bench mark. In that year, local water supplied only 5% of water needs; imported water purchased from MWD supplied the other 95%. Fast forward to 2003. The chart on the following page shows that not much changed in terms of water independence from MWD; it still supplied 85% of the county's water needs. And, significantly, less than half, 48% was firm — that is, protected by SDCWA's preferential rights at MWD. In other words, 37% of the County's water supply was "non-firm," available only at the discretion of the MWD board, a situation that, as earlier discussed, was intolerable to SDCWA.

The 2003 agreements described above will dramatically change that picture. The charts on the following page show the projected result. Imported water purchased from MWD falls from 95% in 1991 and 85% in 2003 to 19-28% in 2020, depending on the amount obtained from desalination, which is shown at 10%-19%. The big difference lies in water obtained from Imperial Valley — a total of 30%, with 21% coming from the agricultural water transfer under the 1988 agreement and 9% from the new canal lining transfer. This huge drop in dependence on MWD — from 85% down to less than 30% — is what SDCWA has been fighting for during the years of disputes with MWD.

This result is certainly cause for celebration in San Diego County, but of course it makes MWD unhappy since it loses very substantial water sales revenue. On the other hand, MWD will receive from San Diego higher "wheeling charges" for transporting the Imperial Valley water to San Diego through its Colorado River Aqueduct. Since those charges start at about $253/acre-foot, this makes up a considerable portion of losses on water sales to San Diego that start at $326/acre-foot.

Returning to the settlement ceremony, Governor Davis in his formal remarks recalled Mark Twain's famous dictum that "whisky is for drinking but water is for fighting," adding, "what we're doing is basically ending years and years of feuding in the West." In addition to the fights among southwestern states over the Colorado River, he might have had in mind MWD's foot-dragging and sometimes outright opposition to the complex of agreements during their nine-year development and particularly during the governor's efforts throughout 2003 to negotiate an end to the disputes.

The long-standing differences on water supply matters between SDCWA and MWD were reflected, perhaps unconsciously, in the remarks of the leaders of those two agencies. Said SDCWA General Manager Maureen Stapleton: "Today we crossed the finish line of a long and arduous marathon." On the other hand, MWD and its president, Ron Gastelum, did not join in the jubilant applause generally given to the final settlement. In his published remarks and choosing words that were considerably less than enthusiastic,

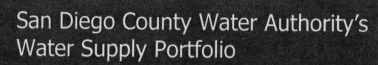

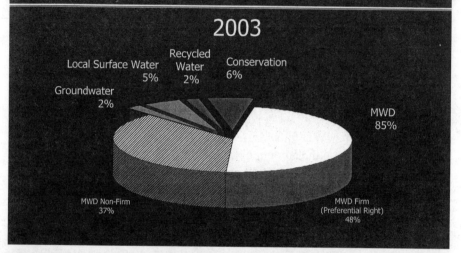

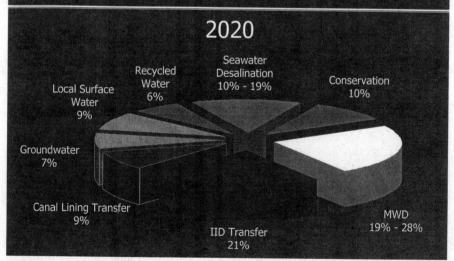

SDCWA water supplies, 2003-2020
San Diego County Water Authority

Gastelum said, "This is a good milestone. Metropolitan is paying $88 million over 15 years to buy a total of 1.6 million acre-feet from Imperial Valley."

Observers called this a disingenuous attempt to use MWD dollars committed by MWD from future Imperial Valley water deals to hide the fact that MWD alone of all the parties avoided any commitment of up-front funds for mitigation of the environmental consequences of taking the Imperial Valley agricultural water. And the *Los Angeles Times* ignored San Diego's accomplishment while predictably giving MWD its support, applauding MWD for "allowing" its Colorado River Aqueduct to be used to carry the water from Imperial Valley, thereby making the extra "wheeling charge" exacted by MWD from SDCWA during the negotiations for carrying the water seem like a gracious gesture.

As the *San Diego Union-Tribune* has editorialized, "there may be nearly as many pitfalls in the future as there were in the past." Discontent coupled with uncertainty still prevails in some sectors of Imperial Valley. It was predicted that the plan to fallow thousands of acres of farmland would upset important agricultural interests in the valley. Events have somewhat ameliorated this concern as a number of Imperial Valley farmers did quickly sign up for the initial land-fallowing program, but opposition to the fallowing program still exists in the valley.

Some skeptics including the *Imperial Valley Press* have argued that "the Salton Sea restoration plan is more rhetorical than real," and implementation of that plan has been complicated by findings that portions of the bottom of the sea are covered with a 50-foot-thick layer of silt having the consistency of peanut butter. Some detractors have deplored the "wiggle room" in the final agreements for the U.S. Department of the Interior to claim in the future as it sometimes has in the past that Imperial Valley's wastage of Colorado River water violates the "reasonable and beneficial use" doctrine imposed by law that "waves a red flag" at Imperial Valley farmers.

In a post-finalization interview, SDCWA Deputy General Manager Dennis Cushman was asked about the possibility that legalities could snarl implementation of the deal; several lawsuits were filed soon after the closing. Cushman responded, "There will be some roadblocks thrown up and some hiccups along the way, but we believe that we have crafted a durable agreement that will withstand the test of time and any legal challenges." Cushman cautioned that lining the All-American and Coachella canals in Imperial Valley is "an enormously complex engineering challenge." These canals carry 3.3 million acre-feet/year — far more than the huge Colorado River Aqueduct.

Cushman noted that further ag-urban water transfers are contemplated by SDCWA long-range water supply planning, but stressed, "seawater desalination is the next major horizon of water opportunity for San Diego County." He added the obvious facts that desalinated water is drought-proof, needs no further treatment, and is in ocean-unlimited supply. As to the future of working with MWD, Cushman was optimistic but remained adamant with respect to the preferential water supply and voting rights issues, the long-lasting disputes raised by SDCWA. "Maureen Stapleton and I agree that these disputes are still alive," he said. "They remain as high-priority, intractable issues that SDCWA will continue to have with MWD."

General Manager Maureen Stapleton looks to the future with the past in mind in meeting the challenges posed by responsibility for a reliable water supply to satisfy San Diego's $126 million economy and three million people. "We have consistently chosen a path of willing buyer/willing seller to purchase our needed water. We have demonstrated our willingness to pay for what we get. I also firmly believe that SDCWA has produced a model for how government should address the existing California imbalance of urban and agricultural water usage."

CONCLUSION

A lone voice entered a dissent to the paeans of praise for the dramatic 2003 Colorado River water settlement. Philip L. Fradkin, author of *A River No More: The Colorado River and the West,* called the agreement a "relatively minor readjustment in the use of Colorado River water," and "less a historic shift than a continuum" of battles. "The reality is that those wars are far from over," he predicts, chiding Interior Secretary Gale Norton with "whistling Dixie" for proclaiming "conflict on the river is stilled." To the contrary, asserted Fradkin. "Traditional interest groups remain in place and are ready to pounce upon each other more ferociously and with greater desperation than before as the Colorado dwindles."

That dwindling of the Colorado comes at a time when drought years continue to threaten water supplies for the entire southwest, as the first chapter of this text records they have periodically done in San Diego. But today, climate experts predict more severe drought than San Diego has ever experienced. Not just "dry times" for a few years as during the 1987-1991 drought, but a decades-long drought. The Jet Propulsion Laboratory in the Pasadena area, recognized for its long-range weather prediction accuracy, speaks of a 20-year drought already well underway. The *New York Times* editorialized dramatically on May 10, 2004: The entire drainage of the Colorado River — the ultimate source of water for much of the metropolitan west — is at risk."

Assistant Secretary Bennett W. Raley warned on December 13, 2003 that the delayed "soft landing," the now-planned gradual reduction in California's over-usage of surplus Colorado River water, could become "a hard landing" as early as the year 2005. The awesome severity of what may well lie ahead was emphasized editorially by the *San Diego Union-Tribune* on May 2, 2004: "We have to start planning for the inevitability of catastrophic drought."

The drought challenge to San Diego's long-term water supply comes just when, as San Diego's *North County Times* put it, San Diego County residents would finally "be able to pour themselves a big glass of history" as the long-awaited Imperial Valley water began to arrive. Anticipating the problem, the new California Water Plan announced in late 2003 by the California Department of Water Resources, the first update in five years, calls for increases in conservation, recycling, increased use of groundwater, and even seeding clouds to generate more precipitation (not mentioning either the Jules Verne fantasy or Charles Hatfield and the 1916 flood). But even with these measures, the plain fact remains that increased ag-urban water transfers will be required. As urban populations grow, southern Californians and their political representatives will look with increasing avidity on the water that experts say is being wasted throughout the vast agricultural areas of the state. More and more people will look with raised eyebrows on the existing imbalance where considerably more Colorado River water flows to plants than to people.

Thus, as the *San Diego Union-Tribune* has editorialized, as drought deepens, the 1988 San Diego-Imperial Valley water transfer agreement may "become a template for future California water transfers." The largest ag-urban transfer in history, it will be looked upon as a model as many more such transfers are fought over in other parts of California where agricultural interests have watched the anguish displayed by the Imperial Valley farmers during the multi-year battle that preceded this agreement.

Central California's Stockton *Record* editorialized strong approval for the agreement but noted local farmers' fear that it is "a huge gamble that threatens to forever change the face of agriculture." The conflicts that increased numbers of these transfers will create in many parts of the state may well escalate beyond those experienced in Imperial Valley. And the hundreds of millions of dollars to be paid for environmental and socioeconomic mitigation measures will almost certainly "up the ante" for ag-urban water transfers in other parts of California.

Add to all of this the perception in northern California that southern California's unsatisfied appetite for more water will cause another clash over the Sacramento River-San Francisco Bay delta. This promises to add California north-south emotional conflict to those that will erupt between agricultural and urban areas in the remainder of the state.

In southern California, MWD's effort to play down the significance of the finalization of the San Diego water transfer agreement by calling it only a "milestone" in comparison to the SDCWA judgment that it was "crossing the finish line," and MWD's reluctance to fully participate in the cost of environmental mitigation measures does not bode particularly well for the future of the relationship between the two umbrella agencies. But this should not be allowed to diminish the fact that SDCWA won a clear victory in its long, long battle to independently assure a large portion of San Diego's future water needs and simultaneously to substantially lessen its reliance on MWD. Indeed, perhaps a measure of the importance of the agreement for San Diego lies in the mild and disingenuous MWD reaction to the event.

Financial burdens will certainly continue to trouble the two umbrella water agencies, and there are other difficult issues. In April 2004, the SDCWA board voted unanimously to take another look at a conveyance system that could bring Colorado River water to San Diego over the mountains, without buying it from MWD. MWD will almost surely oppose any such step by SDCWA. It is still not clear that MWD will fully recognize SDCWA's right to seek a fully diversified water supply, to have solid assurance of its share of Colorado River water from MWD in times of drought, and to secure a stronger vote on the MWD board. The *Los Angeles Times,* which has regularly supported MWD, recently noted editorially even after the highly proclaimed settlement of the IID ag-urban transfer dispute that in a drought situation, "San Diego's portion could be cut back to preserve supplies for other, more senior Metropolitan customers." Thus, the long-lasting preferential water rights and voting power disputes that SDCWA has had with MWD for many years are still very much alive.

MWD, Metropolitan, Mighty Met, or Mother Met, whatever name one uses, has been seen by many to be unduly dedicated to an insistence on being the "boss agency" with unbridled discretionary control over imported water distribution throughout southern California. Los Angeles plays a major role in that agency and this is not the first time that San Diego has been at odds with sister cities to the north. One thinks of young Jim Birch who persisted but ultimately lost in his effort to deliver mail to the entire west coast via San Diego. Admiral Dewey's edict that "San Diego was among the ports of the second order of strategic importance" was ignored by the city fathers and military leaders who made it America's finest military port. San Diego took the Marine and Navy training stations away from San Francisco and fought successfully for decades for its harbor to be home port for many of the Navy's aircraft carriers.

Perhaps General Manager Maureen Stapleton is trying to take SDCWA down the same path as earlier leaders by refusing to accept docilely the dictates and power plays of others and continuing to assert a critical measure of water supply independence through placing San Diego in the forefront of those seeking to solve eternal water problems by even greater emphasis on conservation and water recycling while harnessing the water power of the ocean with new desalination plants, joining the many others already being planned in California.

Surely, all hope that MWD and SDCWA, southern California's two major umbrella water supply agencies, will work out their differences both soon and amicably and that both agencies will work together to provide substantial support for conservation, recycling, ag-urban water transfers and ocean water desalination. Hopefully, too, all San Diego County officials will back up Maureen Stapleton and her SDCWA board on the expensive steps needed to be taken.

Whichever way SDCWA goes in solving the severe water supply problems that lie ahead, taxpayers will have to accept higher price tags for San Diego County's measure of water independence. If San Diego continues to grow as a desirable, attractive place where people can enjoy life, where business and industry can thrive, and where vital military facilities can operate and expand, very substantial funds for water supply will inevitably be required.

Independence never comes cheap. That is, unless wildly improbable annual rainfall far exceeds historical amounts. Asked one San Diego wag, "Anyone for a call to Charles Hatfield the Rainmaker?"

SELECTED REFERENCES

Adams, Austin. *The Story of Water in San Diego*. San Diego, California: California Mountain Water Company, 1909.

Boyle, Robert H., John Greeves, and T.H. Watkins. *The Water Hustlers*. San Francisco, California: The Sierra Club, 1971.

Burzell, Linden. Oral History Program, San Diego Historical Society. Fall 1993, Vol. 39, No. 4.

Ciani, Kyle Emily. *A Passion for Water: Hans Doe and the California Water Industry*. Journal of San Diego History, Spring 1999, Vol. 45, No. 2.

Courtemanche, Carl Joseph. *Utilization of Water in San Diego, 1890-1940*. Master's Thesis, San Diego State University, 1982.

Cranham, Greg T., ed. *Water for Southern California: Water Resources Development*. San Diego Association of Geologists, 1999.

Davis, Edward J.P. *Historical San Diego, Birthplace of California*. San Diego, California: Pioneer Printers, 1953.

Engstrand, Iris H.W. *San Diego: California's Cornerstone*. San Diego, California: Continental Heritage Press, 1980.

Erie, Steven P., and Pascale Joassart-Marcelli. *Unraveling Southern California's Water/Growth Nexus: Metropolitan Water District Policies and Subsidies for Suburban Development, 1928-1996*. California Western Law Review, Spring 2000, Vol. 36, No. 2.

Fletcher, Edward. *Memoirs of Ed Fletcher*. San Diego, California: Pioneer Printers, 1952.

Fowler, Lloyd C. *History of the Dams and Water Supply of Western San Diego County*. Master's Thesis, University of California, 1952.

Golakoff, Ivan, ed. *To Quench a Thirst: Brief History of Water in the San Diego Region*. San Diego County Water Authority publication, 2003.

Griffen, Harry. *Harry Griffen Manuscript*. San Diego, California: Helix Water District.

Hill, Joseph. *Dry Rivers, Damned Rivers and Floods*. Journal of San Diego History, Winter 2003, Vol. 48, No. 1.

Hopkins, H.C. *History of San Diego: Its Pueblo Lands and Water*. San Diego, California: City Printing Co., 1929.

Hundley, Norris, Jr. *The Great Thirst: California and Water.* Berkeley, California: University of California Press.

Jackson, Donald C. *Building the Ultimate Dam.* Kansas City, Kansas: University of Kansas Press, 1995.

Jennings, William H. *Water Lawyer.* Oral History Program, University of California, Los Angeles, 1967, and *Private Memoirs.*

Kern, Philip. *Earthquakes and Faults in San Diego County.* San Diego, California: Pickle Press, 1993.

Nadeau, Remi. *The Water Seekers.* 4th ed. Santa Barbara, California: Crest Publishers, 1997.

Newmark, Harris. *Sixty Years in Southern California.* 4th ed. Los Angeles, California: Zeitlin and Van Bruge, 1970.

Papageorge, Nan Taylor. *Role of the San Diego River in the Development of Mission Valley.* Journal of San Diego History, Spring 1971, Vol. 27, No. 2.

Pourade, Richard F. *City of the Dream.* San Diego, California: Copley Press, 1977.

Pourade, Richard F. *The Explorers.* San Diego, California: Copley Press, 1960.

Pourade, Richard F. *The Glory Years.* San Diego, California: Copley Press, 1964.

Pourade, Richard F. *Gold in the Sun.* San Diego, California: Copley Press, 1965.

Pourade, Richard F. *The Time of the Bells.* San Diego, California: Copley Press, 1961.

Pourade, Richard F. *The Silver Dons.* San Diego, California: Copley Press, 1963.

Pryde, Philip R., ed. *San Diego: An Introduction to the Region.* 3rd ed. Dubuque, Iowa: Kendall/Hunt Publishing, 1992.

Reisner, Marc. *Cadillac Desert.* New York City, New York: Penguin Books, 1993.

Schmoll, M.E., D.J. Young, and D.L. Schug. "34-Mile-Long Tunnel through the Peninsular Ranges of San Diego County" in Stroh, R.T., ed., *Coastal Processes and Engineering Geology of San Diego, California.* San Diego, California: San Diego Association of Geologists, 2001.

Sholders, Mike. *Water Supply Development in San Diego.* Journal of San Diego History, Winter 2002, Vol. 48, No. 1.

Smyth, William E. *History of San Diego, 1542-1908.*

INDEX

Agricultural water wastage 114, 115, 116, 127, 134
Agricultural-urban water transfer: *see* Water transfer
Alamo-Imperial Canal 32, 33, 34
All-American Canal x, 34-38, 39, 49, 50, 116, 122, 133, 146, 147, 151
Alvarado Water Treatment Plant 81, 86
American Institute of Architects 88, 90
Arizona 31, 32, 34, 35, 36, 40, 43-45, 99, 101, 148
Audubon Society 120

Babcock, Elisha 14, 16, 20
Baja California 33, 35, 133
Ball, Harold 130, 135, 137
Barrett Lake, Dam 14, 16, 17, 19, 40, 84, 86, 105
Bass brothers 118, 119
Bay-Delta Action Plan 102, 103
Belock, Frank, Jr. 94, 95, 96
Bi-national canal: *see* Regional Colorado River Conveyance
Black Canyon 40, 41
Bombay Beach 120
Bond issues 22, 23, 66, 75, 82, 131, 136, 138
Bono, Mary 121
Borrego Valley 39, 92
Boulder Dam 36, 39, 40, 42, 43
Brown, Edmund G. (Pat) 70
Burzell, Linden (Lin) 72, 73, 78, 140

CALFED 102, 103
California Aqueduct xi, 56, 57, 58, 59, 69, 70, 103, 114, 116, 127, 145
California Coastal Commission 110
California Department of Water Resources 55, 134, 153
California Development Company 32, 34
California Supreme Court 23, 132
California Water Plan 55, 153
Camp Pendleton 75, 77, 81, 82, 91, 92, 106
Canal lining 116, 122, 146, 149, 151
Capital Improvement Program (CIP) 79, 82, 97
Carlsbad 17, 92, 111
Carlsbad Municipal Water District 77, 82
Central California 55, 56, 154
Central Valley 102, 127
Chaffey, George 32
Chalfant, W.A. 29
Chandler, Harry 33
Chinatown 29
Chocolate Mountains 123
Chollas Reservoir 15
Christian Science Monitor, The 148
Chula Vista 82, 89, 90, 106
"Claim in a can" 35
Coachella Canal, Branch 38, 116, 146, 151
Coachella Valley 38, 123

Coachella Valley Water District 100, 121, 145, 146, 147, 148
Colorado 31, 35, 101
Colorado Desert 32, 119
Colorado River x, 2, 24, 25, 31-48, 49, 51, 52, 57, 60, 61, 65, 67, 69, 70, 71, 72, 73, 75, 79, 81, 88, 91, 97, 99, 100, 101, 103, 110, 116, 117, 118, 119, 122, 126, 129, 133, 134, 135, 138-149, 151, 153, 154, 155
    Surplus water 99, 100, 101, 103, 126, 148
    Water priorities 67, 100, 101, 148
Colorado River Aqueduct (CRA) x, xi, 36, 45-48, 49, 50, 51, 52, 55, 61, 65, 72, 80, 97, 99, 114, 116, 122, 133, 135, 147, 148, 149, 151
Colorado River basin 35, 99, 144, 146, 148
Colorado River Commission 35
Colorado River Compact 35, 42, 99, 101
Columbia River 103
Conservation: *see* Water conservation
Consolidated Water Company 22
Coronado, Water Company 14, 21, 84
Cottonwood Creek 19, 24
Crowe, Frank T. 39, 43
Cunningham, Randy (Duke) 111
Cushman, Dennis 117, 118, 135, 136, 137, 146, 151, 152
Cuyamaca Reservoir, Dam: *see* Lake Cuyamaca
Cuyamaca Mountains 39, 90
Cuyamaca Water Company 20, 21, 22, 24

Davis, Gray 144, 145, 148, 149
Del Mar 8, 17, 19, 77, 82, 84, 91, 94, 95
Delta xi, 56, 102, 103, 127, 144, 145, 154
Desalination: *see* Water desalination
Diamond Valley Lake xi, 98
Discover 120
Diversifying water supplies 116, 129, 130, 135, 136, 137, 155
Doe, Hans 71, 72
Droughts x, 5, 93-94, 97, 101, 113; Drought of 1987-91 xi, 70-71, 93, 104, 106, 115, 130, 153
Dulzura Creek, Conduit 14, 15

Earthquakes 20, 28, 95, 96, 97, 98, 142
East Branch 58, 59, 67, 69, 70, 72, 73
East County 92, 96
Eastwood, John S. 17, 19
Eaton, Fred 27
Economic Study Group 137
Economides, John 97, 139, 140
El Cajon, Valley 4, 14, 21, 22, 87, 88, 90
El Capitan Lake, Dam x, 22-25, 39, 84, 86, 88, 94, 104, 105
El Centro 33, 119, 125
El Monte Tunnel 13

Elsinore fault zone 96
Emergency planning, storage xi, 82, 85, 86, 87, 90, 93, 97, 98
Encina Power Station 111, 112
Endangered Species Act 102
Engstrand, Paul 72, 73, 132, 140
Environmental issues 88, 95, 101, 102, 103, 110, 120, 121, 123, 127, 134, 136, 144, 145, 146, 147, 148, 151, 154
Erie, Steven P. (Steve) 73, 119, 139
Escondido 19, 77, 82, 90, 91, 92, 105, 108

Fallbrook 91, 92
Fallbrook Public Utility District 77, 81, 91
Fallowing 119, 124, 125, 147, 151
Feather River xi, 49, 55, 56, 57, 60, 61, 65, 69, 70, 73, 75, 88, 91, 97, 99, 102, 103
Fletcher, Edward (Ed) 17, 19, 20-23, 24, 88, 104
Fletcher Hills 104
Floods x, 5, 7, 8, 9, 17, 33, 34, 94, 95, 119, 153
Flumes 3, 12, 13, 14, 21, 22, 88
Fogerson, David 142
Forster, John 5
Fradkin, Philip L. 153
Frahm, Christine 73, 130
Freeman, John R. 17

Gardner, Larry 86, 87
Gardner, Michael 134
Gastelum, Ron 112, 125, 132, 134, 135, 136, 139, 142, 144, 146, 148, 149, 151
Griffen, Harry 25, 39, 47, 49, 66, 69-72, 129
Griffen Park (Harry Griffen Park) 88
Grossmont Reservoir 88
Groundwater, facility 75, 79, 86, 90, 91, 92, 104-106, 109, 113, 114, 153
Gulf of California 2, 31

Harper's Weekly 2, 16, 43
Hatfield, Charles, and Hatfield Flood (1916) x, 7, 8, 9, 17, 70, 94, 95, 153, 155
Heilbron, Fred 35, 51, 55, 56, 66, 67, 68, 69, 72, 78
Helix Water District 77, 79, 82, 84, 87, 88, 105
Henshaw Reservoir, Dam: *see* Lake Henshaw
Henshaw, William 17, 20, 24
Hentschke, Dan 136, 137
Higgins, Shelley J. 34, 35
Hill-Ready-Buwalda engineering report 39
Hodges Reservoir, Dam: *see* Lake Hodges
Hollingsworth, Dennis 145
Hoover Dam x, 36, 39-43, 147
Hoover, Herbert 35, 36, 40, 42
Hunter, Duncan 112
Hurricane Kathleen 124

Ickes, Harold 42, 45
Imperial Canal, Dam 32, 36, 37
Imperial County x, 34, 121, 123, 125

Imperial Irrigation District (IID) x, xi, 34, 36, 38, 49, 100, 116, 117, 118, 121, 122, 124, 126, 129, 135, 141-148, 155
Imperial Valley x, xi, 2, 32, 33, 34, 36, 38, 39, 49, 100, 103, 114, 116-127, 130, 133-136, 138, 140, 144-149, 151, 153, 154
Imperial Valley Chamber of Commerce 120
Imperial Valley Press 127, 151
Independence issues, SDCWA: *see under* SDCWA
Infrastructure costs 73, 131, 135, 139
International Wastewater Treatment Plant 108

Jenkins, Logan 142, 143
Jennings Reservoir: *see* Lake Jennings
Jennings, William H. (Bill) 35, 49, 50, 51, 55, 56, 60, 65, 66, 67, 68, 69, 72, 78, 80
Jensen, Joe 66, 67, 69
Johnson, Hiram 35, 36

Kaiser, Henry J. 39
Kimball brothers, company 14, 20, 89
Kimball, Dan 66, 67, 69
Knowland, Senator 67
Knox, Mayor 51
Krauel, Francesca 73, 130

La Mesa 14, 21, 22, 81, 87, 88
La Mesa, Lemon Grove and Spring Valley Irrigation District 22, 23, 88
La Playa 4, 6
Laguna Declaration 68, 71, 131, 140
Lake Cahuilla 33, 119
Lake Cuyamaca, Dam 12, 19, 88, 105
Lake Havasu 45
Lake Henshaw, Dam 19, 90, 91, 105
Lake Hodges, Dam 17-19, 84, 86, 94, 95, 98, 105
Lake Jennings 88, 105
Lake Matthews x, 48
Lake Mead 101, 103
Lake Miramar xi, 84, 86
Lake Morena, Dam 7, 8, 14-16, 19, 40, 84, 86, 105
Lake Murray, Dam xi, 17, 18, 19, 84, 86
Lake Oroville, Dam xi, 56, 144
Lake Perris xi
Lake Powell 101
Lake Shasta 144
Lake Skinner xi, 57, 70, 73, 79, 81, 84, 90, 91, 132
Lake Sutherland 84, 86, 105
Lake Wohlford 105
Lakeside 13, 22, 88, 89, 90
Landers earthquake (1992) 96
Last Oasis 106
Leach, Mike 130
Leimgruber, Wally 117, 125
Lemon Grove 22, 87
Levy, R.M., Treatment Plant 88, 89
Long-Range Water Resources Plan (2002) 86

Los Angeles Aqueduct x, 28
Los Angeles control over MWD: *see under* MWD
    see also San Diego and Los Angeles discord
Los Angeles County 43, 69
Los Angeles Times 28, 33, 34, 35, 43, 56,
    144, 151, 155
Los Angeles Water Company 27
Los Coches Trestle 12
Loveland, George 130, 135, 136, 137
Loveland Reservoir 89, 105
Lower Basin states 31, 35, 99, 100, 101
Lower Otay Lake, Dam: *see* Otay Lakes

Madigan, Mike 71, 73, 79, 130, 133, 134, 137
Mason, Dale 140
McClary, Michael 110
McKinnon, Clinton 67
Metropolitan Water District (MWD) x, 43, 45, 46,
    49-52, 56-60, 61-73, 78, 79, 81, 84, 91, 93, 97,
    98, 99, 100, 101, 102, 103, 110, 112, 114, 116-
    118, 119, 122, 125, 126, 129-152, 154, 155
    MWD, Los Angeles control over 65, 66, 71, 73
    MWD member agencies 62, 63, 64
    MWD opposition: *see under* Water transfer
    MWD voting rights 64
    MWD water plans 93, 139, 140
Mexico 1, 4, 5, 23, 25, 31, 32, 34, 94, 101, 133
Miller, Joaquin 7
Miramar Reservoir: *see* Lake Miramar
Miramar Water Treatment Plant 86
Mission Basin, Desalting Facility 105, 109
Mission Beach 21
Mission Dam: *see* Old Mission Dam
Mission Gorge 3
Mission San Diego de Alcalá 3
Mission Valley 6, 8, 11, 94, 95, 104, 109
Moeur, Benjamin B. 43, 44, 45
Mojave Desert 27, 28
Morena Reservoir, Dam: *see* Lake Morena
Morena-to-Otay-to-Chollas system 22
Morgan, Michael 118
Mount Olympus 133
Mount San Jacinto (Old San Jack) tunnel 47, 48,
    52, 80
Mountain Water Company 14, 16, 20, 21, 22, 24
Mulholland, Catherine 29
Mulholland, William 27, 28, 29, 65
Murphy, Mayor 73, 108
Murray, James 17
Murray Reservoir, Dam: *see* Lake Murray

National City, Well Field 77, 82, 84, 86, 89, 106
Nevada 31, 35, 39, 40, 99, 101, 103, 145, 148
New Mexico 31, 35, 101
New York Times 123, 148, 153
Niver, Norman 121, 123
Nixon, Senator 67
North American Water and Power Alliance
    (NAWAPA) 103

North City Water Reclamation Plant 108
North County 17, 20, 72, 86, 91, 94, 108, 137
North County Times 148, 153
Northern California 55, 103, 144, 145, 154
Norton, Gale 147, 148, 153

Oceanside 8, 77, 82, 91, 92, 105, 106, 109
Old Mission Dam, Flume x, 3, 4
Old Town San Diego 6
Olivenhain Municipal Water District 77, 82, 92
Olivenhain Reservoir, Dam xi, 98
Operational Area Emergency Plan 97
Orange County 43, 69, 112, 134
Orchid Awards 88, 90
Oregon 103
Oroville Dam: *see* Lake Oroville
Otay Lakes, Dams xi, 8, 15, 16, 17, 19, 40, 80, 84,
    86, 105
Otay River, Mesa 7, 19, 24, 108
Otay Water District 77, 82, 84, 92, 108
Otay Water Treatment Plant 86
Otis, Harrison Gray 33
Owens, Bill 101
Owens River, Lake 27, 28, 61, 65, 70, 123
Owens Valley x, 27, 28, 29, 55, 103, 117, 123

Pace, Philip 131, 134, 144
Padre Dam Municipal Water District 77, 82, 84,
    90, 108
Padre Dam Water Recycling Facility 90
Palm Springs 38
Palo Verde Irrigation District 100, 116
Parker Dam x, 36, 39, 43, 44, 45, 61
Parker, George 132
Pauma Valley 91
Perdue, Robert A., Water Treatment Plant 90
Peripheral Canal xi, 56
Peterson, John 92
Pipelines xi, 11, 14, 49, 51, 52, 56, 57, 60, 65, 66,
    67, 69, 79, 80, 81, 82, 86, 88, 90, 91, 96, 97, 98,
    109, 111, 132, 133, 139
Point Loma 4, 6
Point Loma desalination plant 110, 111
Point Loma sewage treatment station 96
Popular Science 38
Poseidon Resources 111
Postel, Sandra 106
Poway 77, 82, 92, 108
Powell, John Wesley 31
Preferential rights: *see under* Water rights
Priorities, water: *see under* Colorado River
Propositions 66, 68, 144, 121
Pryde, Philip R. 11, 65, 136
Public Works Administration 36
Purple pipe system, projects 107, 108, 109

Qualcomm Stadium 97, 109
Quantification Settlement Agreement (QSA) xi,
    102, 126, 134, 141, 143, 144, 145, 146, 147

Rainbow Municipal Water District　77, 82, 84, 91
Rainfall　2, 3, 4, 82, 84, 88, 104; 100-year　94-95;
　"Miracle March" (1991)　70-71
Raley, Bennett W.　153
Ramona Lake　105
Ramona Municipal Water District　77, 82, 92, 105
Rancho Santa Fe　17
Reclamation and recycling: *see* Water reclamation
　and recycling
Regional approach　71, 72, 73, 129, 130, 139, 140
Regional Colorado River Conveyance　133
Regional Urban Water Management Plan
　(MWD 2000)　140
Regional Water Facilities Master Plan
　(SDCWA 2002)　111, 113, 114
Reservoir water usage　82, 84, 85, 86, 87, 89, 105
Reverse osmosis desalination　90, 109, 110
Reynolds, Richard A., Groundwater Demineraliza-
　tion Facility　90, 106
Rhinerson, Bernie　117, 147
Rincon del Diablo Municipal Water District　77, 82,
　91, 92
Riparian (River) rights: *see under* Water rights
River No More: The Colorado River and the West, A
　153
Riverside County　x, xi, 43, 47, 52, 81, 121
Roberts, Paula　82
Rodriquez Dam　95
Roosevelt, Franklin D.　36, 50
Rose Canyon fault zone　96
Roseville　6
Rossman, Antonio　123

Sacramento, Valley　55, 118, 144, 145
Sacramento River　56, 102; *see also* Delta
Saint Francis Dam　28
Salton Sea　x, 34, 38, 116, 118-125, 126, 127, 144,
　147, 148, 151
Salton Sea Authority　121, 125, 147
Salton Sea Reclamation Act　121
Salton Sink, City　33, 120
San Andreas fault, zone　96, 97, 98
San Bernardino County　43
San Diego & Arizona Eastern (SD&AE) Railroad
　21, 22
San Diego and Los Angeles discord　52, 56, 65, 66,
　67, 68, 69, 71, 72, 73, 131, 132, 138, 155
San Diego Aqueducts　x, xi, 50, 52-55, 66, 80, 81,
　98, 122
San Diego Bay, Harbor　1, 3, 72
San Diego City Lakes (reservoirs)　84-87, 99, 105
San Diego, City of, Water Department　70, 77, 84-87
San Diego County reservoirs　104, 105
San Diego County Water Authority (SDCWA)　x,
　xi, 51, 52, 60, 61, 65-73, 75-84, 86, 88, 90, 91,
　92, 93, 94, 97, 98, 99, 102, 103, 104, 105-109,
　111-114, 116, 117, 118, 119, 122, 125, 126, 127,
　129-152, 154, 155

SDCWA annexation by MWD　51, 52, 65
SDCWA independence issues　50, 60, 70, 71-73,
　129, 130, 134, 135, 136, 137, 149, 154, 155
SDCWA member agencies　76, 77
SDCWA pipelines　79-82
SDCWA preferential rights　114
SDCWA voting rights　77
SDCWA voucher programs　106, 107
SDCWA water demand, supply (charts, tables)
　83, 113, 126, 143, 150
SDCWA water plans　93, 94, 111, 132, 137, 139
　(tables)　107, 113, 114
San Diego Flume, Company　12, 13, 14, 21, 22, 88
San Diego Formation　90, 104, 105
San Diego River　3, 6, 7, 8, 14, 19, 22, 23, 24, 86,
　90, 94, 95
San Diego Union, Union-Tribune　12, 14, 20, 21,
　34, 39, 134, 142, 146, 151, 153, 154
San Diego Water Company　11, 14, 22
San Dieguito Mutual Water Company　17
San Dieguito River　7, 17, 19, 24, 94, 95
San Dieguito Water District　77, 84, 92
San Francisco Bay　56, 102; *see also* Delta
San Jacinto fault zone　96
San Joaquin River　56, 102; *see also* Delta
San Luis Rey River　7, 17, 19, 90, 91, 94, 95, 105,
　109
San Onofre Nuclear Generating Station　110, 111
San Pasqual Valley　104
San Vicente Lake, Dam　x, xi, 50, 52, 81, 84, 86,
　94, 98, 105, 133
San Vicente Pipeline　98
Santa Fe Irrigation District　77, 84, 92
Santa Fe Railroad　17
Santa Margarita River　7
Santee Lakes Regional Park　90, 108, 109
Savage, Hiram N.　17, 19
Scattergood plant　112
Section 135　67, 68, 73, 93, 114, 131, 132, 138, 140
Seven-Party Agreement　100, 101
Sholders, O.B. (Mike)　95
Sierra Nevada　27
Six Companies, Inc.　39, 43, 44, 45
Smithsonian　27, 29
Snake River　103
Snow, Lester　130
South Bay Irrigation District　77, 82
South Bay Power Plant　111
South Bay Water District　89
South Bay Water Reclamation Plant　108
South County　14, 86, 94, 95, 108
Southern California　xi, 32, 39, 47, 55, 56, 61, 65,
　71, 72, 93, 98, 99, 100, 103, 110, 115, 118, 126,
　129, 134-146, 154, 155
Southern California Mountain Water Company　14,
　16, 20, 21, 22, 24
Southern Nevada Water Authority　145
Southern Pacific Railroad　33, 34

Spreckels, John D. 14, 20, 21, 22, 24
Spring Valley 22, 87, 90
Stapleton, Maureen 78, 127, 130, 131, 136, 137, 146, 149, 152, 155
State Water Authority, Act 71, 78
State Water Project (SWP) xi, 56, 57, 93, 102, 103
Steele, Kenneth 140
Stockton *Record* 154
Story of Inyo, The 29
Surface water, runoff 2, 82, 92, 94, 104, 113, 114, 122, 126
Sutherland Reservoir: *see* Lake Sutherland
Sweetwater Authority 77, 82, 84, 89, 90, 105, 106, 108
Sweetwater Company 20, 22
Sweetwater Reservoir, Dam 9, 14, 19, 40, 89, 105
Sweetwater River, Trestle 13, 14, 19, 90
Swing, Phil 34, 36, 40, 50
Swing-Johnson Bill 35, 36

Tampa Bay desalination plant 111
Taylor, James 118
Tehachapi Mountains 56, 69
Terrorism 93, 95, 96, 97, 98, 142
Tierney, J.G. and P.W. 43
Tijuana River 7, 14, 19, 94, 95, 108
Torrey Pines Golf Course 108
Town that Launders its Water, The 90
Treaty of Guadalupe Hidalgo 5, 23
Twin Oaks Valley 133

U.S. Bureau of Reclamation x, 39, 43, 50, 66, 101, 102, 106, 121
U.S. Department of the Interior 31, 35, 39, 40, 42, 45, 49, 50, 61, 99, 100, 101, 102, 106, 110, 115, 121, 126, 141, 142, 144, 145, 146, 147, 148, 151, 153
U.S. Desalination Coalition 112
U.S. Environmental Protection Agency 123, 134
Underground storage 105
University Heights Reservoir 7, 15
Upper Basin states 31, 34, 35, 99, 101
Upper Otay Lake, Dam: *see* Otay Lakes
Urban Water Management Plan (SDCWA 2000) 93, 94, 107, 113, 114, 132, 137, 139
Utah 31, 35, 101

Vallecitos Water District 77, 91, 92, 140, 142
Valley Center Municipal Water District 77, 79, 82, 92
Vista Irrigation District 77, 82, 84, 90, 91, 105
Voting rights: *see under* MWD; SDCWA; Water rights

"Wake-up call" (Drought of 1987-91) 70-71, 115
   see also Drought of 1987-91, *under* Droughts
Wastewater 107, 108, 114, 115, 116, 122, 127, 134
Water 2025 (U.S. Bureau of Reclamation) 101
Water boys 6, 11

Water conservation 71, 86, 88, 104, 106, 107, 112, 113, 114, 115, 116, 117, 119, 121, 122, 124, 134, 138, 141, 147, 153, 155
Water demand and supply 80, 93, 101, 104, 113, 114, 126, 127, 129, 130, 134, 138, 143, 149
   (SDCWA charts, tables) 83, 113, 126, 143, 150
Water desalination, plants 90, 109-112, 113, 121, 134, 149, 151, 155; Desalination schematic 111
Water hustlers 11, 19, 20; *Water Hustlers, The* 25
Water Odyssey: The Story of Metropolitan Water District 117
Water plans 5, 55, 86, 93, 94, 97, 102, 103, 107, 111, 113, 114, 132, 137, 139, 140, 153
Water pricing disputes 67-69
Water priorities: *see under* Colorado River
Water reclamation and recycling, plants, projects 86, 90, 104, 107-109, 113, 114, 134, 153, 155
Water rights 5, 23, 27, 43, 52, 75, 90, 99, 100, 114, 118, 129, 131, 155
   Preferential rights 65, 67, 68, 71, 78, 93, 114, 125, 131, 132, 137, 138, 143, 149, 152, 155
   "Pueblo rights" 5, 23   "Plan of Pitic" 5
   "Reasonable and beneficial use" 3, 5, 115, 126, 127, 151
   Riparian (River) rights 5, 17, 23
     "First in time, first in right" 5, 23
     "Law of the river" 5, 23
   Voting rights, powers 61, 64, 65, 66, 67, 68, 73, 77, 78, 79, 82, 131, 132, 138, 152, 155
Water Surplus and Drought Management Plan (MWD 1999) 93, 139
Water transfer, agreement xi, 115-119, 121-127, 129, 130, 133, 134, 135, 136, 138, 139, 140, 141, 142, 143-149, 151, 153, 154, 155
   MWD opposition 118, 134, 136, 138, 143, 144, 145, 146, 154
Water treatment plants xi, 79, 81, 84, 86, 88, 89, 90, 94, 108, 132
Water wells 6, 7, 11, 90, 92, 104, 105, 106, 116
Water: *see also* Groundwater, Reservoir water usage, Surface water
Webb, Walter Prescott 2
West Basin Municipal Water District plant 112
"Wet water" 143
Wheeler, Mark 27
Wheeling charge 116, 133, 135, 147, 149
Whipple Mountains 43, 45
White, Theodore 43
Wilbur, Ray Lyman 40
Wilson, Pete 73
Wohlford Reservoir: *see* Lake Wohlford
World War II, era ix, 25, 46, 49, 65, 93, 120, 133
"World's Biggest Ditch, The" 38
Wunderly, Murray 92
Wyoming 31, 35

Yuima Municipal Water District 77, 79, 84, 91
Yuma Projection California 100

RECOMMENDED READING —
BACKGROUND TO THE SAN DIEGO WATER STORY

Water for Southern California (SDAG 1999)	Greg T. Cranham, ed.
California Desert Miracle: The Fight for Desert Parks and Wilderness	Frank Wheat
The Rise and Fall of San Diego: 150 Million Years of History	Patrick L. Abbott
Coastal Geology of San Diego (SDAG 2001)	Robert C. Stroh, ed.
Geology of San Diego: Journeys Through Time	Clifford, Bergen, Spear
San Diego: An Introduction to the Region	Philip R. Pryde
Thirst for Independence: The San Diego Water Story	Dan Walker
Water and the Shaping of California (Heyday Books)	Sue McClurg
The Water Seekers (Crest Publishers)	Remi Nadeau
Cadillac Desert: The American West and its Disappearing Water (Penguin Books)	Marc Reisner
A River No More: The Colorado River and the West (University of Arizona Press)	Philip L. Fradkin
Water and the California Dream: Choices for the New Millenium (Sierra Club Books)	David Carle

SUNBELT PUBLICATIONS
www.sunbeltbooks.com

Incorporated in 1988 with roots in publishing since 1973, Sunbelt produces and distributes publications about "Adventures in Natural History and Cultural Heritage in the Californias." These include natural science and outdoor guidebooks, regional histories and reference books, multi-language pictorials, and stories that celebrate the land and its people. Sunbelt books help to discover and conserve the natural and historical heritage of unique regions on the frontiers of adventure and learning.